# Studies in Computational Intelligence

Volume 755

**Series Editor**

Janusz Kacprzyk, Polish Academy of Sciences, Warsaw, Poland

The series "Studies in Computational Intelligence" (SCI) publishes new developments and advances in the various areas of computational intelligence—quickly and with a high quality. The intent is to cover the theory, applications, and design methods of computational intelligence, as embedded in the fields of engineering, computer science, physics and life sciences, as well as the methodologies behind them. The series contains monographs, lecture notes and edited volumes in computational intelligence spanning the areas of neural networks, connectionist systems, genetic algorithms, evolutionary computation, artificial intelligence, cellular automata, self-organizing systems, soft computing, fuzzy systems, and hybrid intelligent systems. Of particular value to both the contributors and the readership are the short publication timeframe and the world-wide distribution, which enable both wide and rapid dissemination of research output.

The books of this series are submitted to indexing to Web of Science, EI-Compendex, DBLP, SCOPUS, Google Scholar and Springerlink.

More information about this series at http://www.springer.com/series/7092

Aniket Roy · Rahul Dixit · Ruchira Naskar ·
Rajat Subhra Chakraborty

# Digital Image Forensics

## Theory and Implementation

Springer

Aniket Roy
Department of Computer Science
and Engineering
Indian Institute of Technology Kharagpur
Kharagpur, West Bengal, India

Rahul Dixit
Department of Computer Science
and Engineering
National Institute of Technology Rourkela
Rourkela, Odisha, India

Ruchira Naskar
Department of Computer Science
and Engineering
National Institute of Technology Rourkela
Rourkela, Odisha, India

Rajat Subhra Chakraborty
Department of Computer Science
and Engineering
Indian Institute of Technology Kharagpur
Kharagpur, West Bengal, India

ISSN 1860-949X ISSN 1860-9503 (electronic)
Studies in Computational Intelligence
ISBN 978-981-10-7643-5 ISBN 978-981-10-7644-2 (eBook)
https://doi.org/10.1007/978-981-10-7644-2



This Springer imprint is published by the registered company Springer Nature Singapore Pte Ltd.
The registered company address is: 152 Beach Road, #21-01/04 Gateway East, Singapore 189721, Singapore

# Preface

In today's cyberworld, digital images and videos constitute the most frequently transmitted forms of information. This has been made possible by the great proliferation of low-cost, easy-to-use, and efficient consumer electronic devices such as high-resolution digital cameras for image acquisition and simultaneous availability of high-speed transmission media such as the Internet.

However, the present-day easy availability of low-cost or free image and video processing software tools, often easy to use and having an immense number of multimedia manipulating features, poses a threat to the fidelity of digital multimedia data. Common image and video processing operations such as cropping, splicing, and blurring can be performed at the click of a mouse by a novice user using such software. This situation compels us to question the trustworthiness of digital images and videos. Now-a-days, digital images and videos act as the major electronic evidences for law enforcement across the world, as they are the most effective and efficient means to collect digital evidences from crime scenes. Hence, the maintenance of their trustworthiness and reliability is a major challenge in today's digital world. It is extremely crucial to preserve such digital evidences against cybercrime for a suitable presentation in court of law. The need for the investigation and maintenance of the fidelity and reliability of digital images and videos has given rise to the field of "digital multimedia forensics". Digital image forensics deals with mainly two broad problems: (a) camera source identification and (b) image forgery detection. Camera source identification is the study to analyze which camera model or class has been used to capture a particular image. On the other hand, image forgery detection is applied to identify whether a given image is forged or not; moreover, if yes, it also attempts to localize the forgery. Image forgery detection is inherently a "blind" technique, whereby the original (unmodified) image is not present during the detection process.

The recent surge of interest in this field is also reflected by the fact that the forthcoming (at the time of writing) *IEEE Signal Processing Cup 2018* is focused on the problem of camera source identification. In this book, we focus on the basic techniques used in state-of-the-art research on digital image forensics. Recent advancements in several aspects of this emerging field have been discussed lucidly with illustrations and necessary background in each chapter in the book. The book thus presents a diverse range of multimedia forgery detection techniques based on digital forensics, along with pointers to open research problems and challenges in this field. It is aimed at catering to graduate students, researchers, and practitioners in the area of digital image forensics. We hope this effort can benefit both the researchers from academia and industry professionals working in this or even related domains. We will consider our effort to be successful if through reading this book the readers get interested in this exciting research topic with vast promising possibilities and contribute to the innovations in this budding field.

This book aims to present to its readers, a thorough understanding of state-of-the-art digital forensic techniques for the detection of multimedia forgery. The goal is to equip the reader with the necessary knowledge base about the existing repertoire of techniques, as well as open problems, issues, and challenges faced by a digital image forensic researcher or analyst. The chapter-wise organization of the book is provided below.

- Chapter 1: The importance and relevance of digital image forensics in the context of today's cyberworld have been discussed in the introductory chapter with instances and illustrations.
- Chapter 2: In this chapter, the source identification problem is investigated with a detailed state-of-the-art literature survey. Generally, machine learning-based model building with appropriate features has been proven to be effective for such problem. In this chapter, an efficient solution using *discrete cosine transform residual* feature and ensemble classifier has been discussed in detail with experimental validation.
- Chapter 3: Among all classes of digital image forgeries, copy-move forgery is the simplest yet most widely launched attack to forge an image. In this context, here we present a three-way classification of state-of-the-art digital forensic techniques for copy-move forgery detection or region duplication detection, along with a complete survey of their operating principles. In addition, we analyze the schemes and evaluate and compare their performances in terms of a conventional set of performance metrics, which may be used as a standard benchmark for evaluating the efficiency of any general copy-move forgery detection technique for digital images. The comparison results provided by us would help a user to select the most optimal forgery detection technique, depending on his requirements.
- Chapter 4: *Mean* and *variance* are two widely adopted statistical features of an image. The contribution of an individual image block (with respect to pixel intensity of the entire image) can be captured using *mean*, and *variance* can be used to find how each pixel varies from its neighbors in a block. In this chapter, we present a copy-move forgery detection technique, exploiting mean and variance. Additionally, we present a false-positive optimization technique in copy-move forgery detection, based on the variance of image blocks.

- Chapter 5: Most state-of-the-art copy-move forgery detection techniques incur large false-positive rates for the special case of detecting forgeries involving similar but genuine objects. In this chapter, that particular and unexplored problem has been addressed with *rotated local binary pattern*-based textured features.
- Chapter 6: In this chapter, we present a region duplication detection technique, operational in the transform domain, which utilizes the *undecimated dyadic wavelet transform* (DyWT) of an image. An image is divided into pixel sub-matrices or *blocks*, and matches among different image blocks as found, so as to detect image region duplications. *Similarity* between different blocks, with respect to features extracted using DyWT, is computed. Subsequently, an eight-connected neighborhood check technique is adopted to reduce false positives.
- Chapter 7: We conclude the book with open problems and directions for future research.

We have tried our best to minimize mistakes and typographical errors in the document; however, we would thankfully welcome suggestions and corrections, if any, to us via e-mails.

Kharagpur, India — Aniket Roy
Rourkela, India — Rahul Dixit
Rourkela, India — Ruchira Naskar
Kharagpur, India — Rajat Subhra Chakraborty
February 2019

# Acknowledgements

Aniket Roy would like to thank his colleagues at the Secured Embedded Architecture Laboratory (SEAL), Department of Computer Science and Engineering, IIT Kharagpur, for their enjoyable company, helpful assistance, and constant encouragement. He is also grateful to his parents and brother for their blessings and unwavering support.

Rahul Dixit and Ruchira Naskar would like to express their sincere gratitude toward all members of the Department of Computer Science and Engineering, NIT Rourkela, for their cooperation and support.

Rajat Subhra Chakraborty would like to express his gratitude to his family members for their understanding and support when the book was being written and to the members of the Department of Computer Science and Engineering, IIT Kharagpur, for their encouragement.

# Contents

# About the Authors

**Aniket Roy** is an M.S. (research) student at the Department of Computer Science and Engineering at the Indian Institute of Technology Kharagpur. He received his B.Tech. degree in Electronics and Communication Engineering from West Bengal University of Technology (now Maulana Abul Kalam Azad University of Technology), India in 2014. His primary research interest lies in multimedia security, reversible watermarking and digital forensics. He received the Best Paper Award at the 15th International Workshop on Digital-forensics and Watermarking. He is an IEEE student member.

**Rahul Dixit** is a Ph.D. scholar at the Department of Computer Science and Engineering at the National Institute of Technology, Rourkela, India. He received his M.Tech. and B.Tech. degrees in Computer Science and Engineering from the Indian School of Mines, Dhanbad and Uttar Pradesh Technical University (now Dr. A.P.J. Abdul Kalam Technical University), respectively. His major research interests include digital image and video forensics, multimedia security and image processing.

**Ruchira Naskar** has been an assistant professor at the National Institute of Technology Department of Computer Science and Engineering, Rourkela, India since 2014. She received her Ph.D. from the Indian Institute of Technology Kharagpur, India in 2014. Her primary research interests are multimedia security and digital rights management, and she has over 30 publications in reputed journals and conferences. Her recent research interest is Digital Forensics. Dr. Naskar is a Member of the IEEE.

**Rajat Subhra Chakraborty** is an associate professor at the Indian Institute of Technology Kharagpur Department of Computer Science and Engineering, India. He has worked at National Semiconductor and Advanced Micro Devices (AMD). His research interests are in the areas of hardware security, very-large-scale integration (VLSI) design, digital watermarking and digital image forensics. He has published over 80 papers in respected international journals and conferences, and holds two U.S. patents. He has a Ph.D. in Computer Engineering from Case Western Reserve University (U.S.A.). Dr. Chakraborty is a senior member of the IEEE and the ACM.

# Chapter 1
# Introduction

## 1.1 Threats to the Integrity of Digital Media Content

Digital images constitute a significant part of the information disseminated in our usual quotidian communications as well, e.g., social media, websites, newspapers, television, and magazines. However, due to the glut of digital images promulgated on a day-to-day basis over public and easy-to-access communication channels such as the Internet, they are highly vulnerable to alteration and integrity attacks. Additionally, the easy availability of low cost, handy yet powerful image editing software tools, e.g., Adobe Photoshop, Pixar, etc., has put the authenticity and integrity of digital images further at stake nowadays. The impeccable and imperceptible manipulation of the digital images by these software tools are hard to identify by trivial means. Tampering to an image may not always be performed with unscrupulous intentions—image editing may be done with benign intentions as often required for improving the style and visual quality with splendid effects. However, forgers often try to exploit the underlying image processing tools and techniques with an aim to conceal or manipulate salient elements within an image, and subsequently use the tampered image for the purpose of misconstruing information, or to besmirch the reputation of one or more individuals. Because of the proliferation of social networking and instant messaging services, fake or misrepresented information can spread very fast and can have disastrous consequences, potentially at a national scale, far beyond personal reputation loss. In recent years, digital images not only act as prime carriers of information, but they also play an important role act as a form of evidence in a court of law. Hence, protecting the authenticity and integrity of digital images has become a major challenge of late.

As a motivational example, Fig. 1.1 shows several instances of image forgeries, where several images and their fake versions are shown side-by-side. Figure 1.1a and c show face morphing attacks which can be used for personal identity infringement. Figure 1.1b shows former U.S. President Barack Obama watching Indian Prime Minister Narendra Modi's lecture on TV in the fake image [4]. This image was also trending once on Twitter and finally proven to be fake by the forensic analysts.



A. Roy et al., *Digital Image Forensics*, Studies in Computational Intelligence 755, https://doi.org/10.1007/978-981-10-7644-2_1

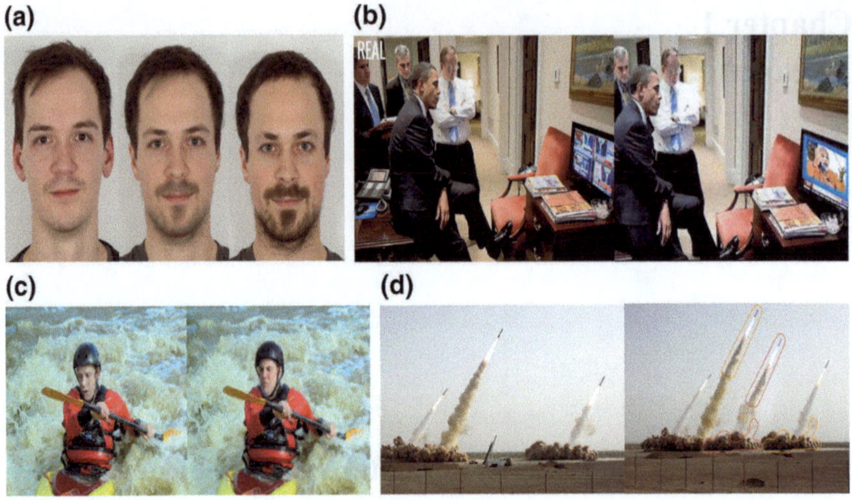


**Fig. 1.1** Image forgery attack examples: **a** face morphing attack: left, right images are original, and center image is morphed. **b** Original image (left), forged image (right) [4]; **c** an original image (left) and a composite image (right), in which the head of another person was overlaid onto the shoulders of the original kayaker; **d** an image of an Iranian missile test taken in July 2008: original (left) and forged (right) [13]

Figure 1.1d shows both original and forged images (with more missiles reported in the forged image) of an Iranian missile test taken in July 2008 [13].

## 1.2 Digital Content Protection

Due to the recent rapid increase in cybercrime rates [5], tremendous growth of research interest has been observed in the fields of digital content protection and digital rights management [2, 6–10] in last couple of decades. The most popular practices in this field environs digital content protection techniques such as *watermarking* and *steganography* [2], which generally perform some sort of pre-processing on the data to make it secured, such as hash computation of cover media and embed it as watermark or embedding some other form of security sensitive data into the cover media, etc. The precomputed data is used a posteriori, for verification, authentication, and privacy protection of the digital media. Such techniques are referred to as *active* cybersecurity measures. Active security measures require some additional auxiliary information for authenticity verification of digital media. Therefore, such auxiliary information is obtained in the pre-processing steps. For example, in digital watermarking, generally the watermark is generated by computing the hash of the cover image and used as auxiliary information. On the other hand, in steganography, the cover media itself is considered as auxiliary information, into which the sensitive image is hidden. Securing multimedia content by steganography or digital watermarking can be enforced by means of special embedded hardware or software into the devices used for image or video capture and acquisition. Nowadays, most of the image capturing devices contain modules for providing such security capabilities such as watermarking, encryption, and data embedding. However, these added security features also enhance the production cost of such devices, which hampers the business of the camera manufacturing companies, since their target being designing and manufacturing cameras with relatively low price for capturing consumer electronics mass market. This practical issue, along with the current surge in cybercrime rates, has engendered the need for blind security measures, which do not require any additional information for verifying the authenticity of digital media.

## 1.3 Digital Forensics

Digital forensics is a developing research domain which investigates the clues and evidences residing in the digital media to identify the origin of the media and also whether it has been tampered with or pristine in nature. Digital forensic techniques neither use any a priori information about that media under inspection nor carry out any pre-computation on the data prior to detect any forgery it has been undergone. The investigations and analyses are totally based on post-processing of data. Hence, digital forensic techniques are considered as *passive* [1] security measures for digital content protection. Naturally, these techniques are more practical, powerful, and challenging. The fundamental differences among the content protection techniques and digital forensics have been depicted in block diagram in Fig. 1.2.

Currently, digital image forensics mainly deals with two major classes of problems and their solutions:

- Image source identification and
- Image forgery detection.

In this book, these problems are described in detail, along with several state-of-the-art approaches for solving them. Next, the individual problems are described with examples.

### *1.3.1 Image Source Identification*

Image source identification deals with the forensic analysis to investigate which device (or class of device) captured or formed the image under investigation. Figure 1.3 refers to the problem, where given a photograph, the forensics problem is to determine which camera was used to capture the photo. Generally, effective feature extraction and classifier learning-based approach are the current solutions to the problem. In the subsequent chapters, several state-of-the-art techniques have been reviewed in detail.

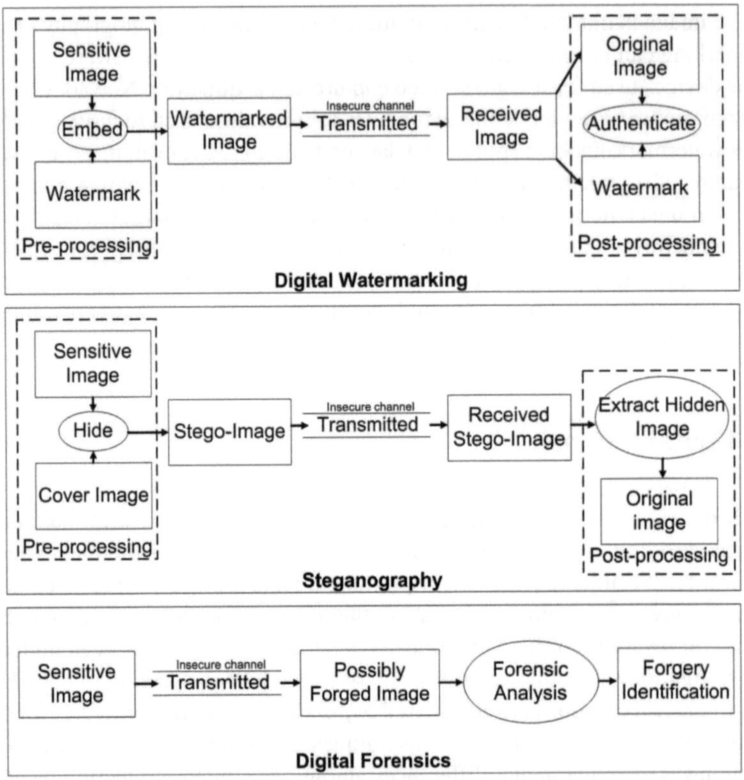


**Fig. 1.2** Block diagrams representing the major operational steps of non-blind digital content protection techniques: (top) digital watermarking, (middle) steganography, and (bottom) blind digital forensic technique

## *1.3.2 Image Forgery Detection*

Image forgery detection is carried out with the aim of investigating whether the image under question represents the unmodified captured scene, or has it been forged to deceive the viewer. There are several classes of image forgeries, which will be discussed hereafter.

### 1.3.2.1 Copy-Move Image Forgery

In copy-move forgery, some part of the image is copied and moved to some other part of the same image. Therefore, the source of the forgery remains within the image itself. Hence, feature matching-based technique is quite useful for the detection of such forgeries. An example is given in Fig. 1.4. This is the most common and widely used type of image forgery.

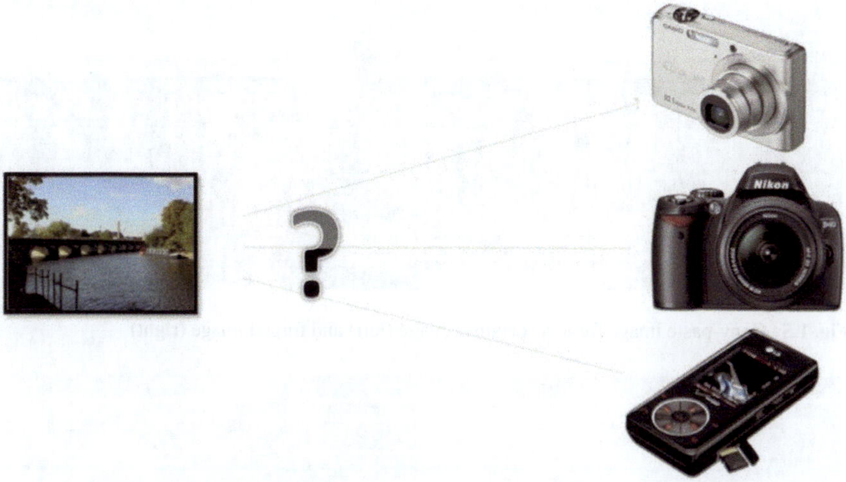


**Fig. 1.3** Camera source identification problem

**Fig. 1.4** Copy-move image forgery

#### 1.3.2.2 Copy-Paste Image Forgery

In copy-paste forgery, some different image is copied and pasted on the image to create the forgery. The major difference of this type of forgery with respect to copy-move forgery is that the source of the forgery is not residing within the image itself. An example is shown in Fig. 1.5. This is a more realistic form of image forgery. Simple feature matching-based detection techniques usually do not work in this case. Therefore, more sophisticated feature- or block-based techniques are usually applied to identify these forgeries.

**Fig. 1.5** Copy-paste image forgery: original image (left) and forged image (right)

**Fig. 1.6** Recompression image forgery: original image (left) and JPEG compressed version (right)

#### 1.3.2.3 Compression and Recompression-Based Image Forgery

When an image undergoes JPEG compression, it exhibits certain compression artifacts in the image. While applying several post-processing operations, it would additionally get compressed [12]. Therefore, such compression artifacts can be treated as forgery. Multiple compressions can seriously downgrade the quality of the image. An example of recompression forgery is shown in Fig. 1.6.

#### 1.3.2.4 Filtering Forgery

In the literature on digital image forensics, there exist many works on copy-move and copy-paste forgery detection and localization approaches. However, comparatively less work has been done for image filtering forgery detection. Generally, forgeries are followed by an enhancement technique to make the forgery more convincing

**Fig. 1.7** Filtering and splicing forgery: original image (left), median-filtered version (middle), and face spliced image (right)

and less detectable. The enhancement techniques might include noise reduction, filtering, contrast enhancement, deblurring, edge sharpening, etc., but among these techniques filtering is the most common and widely used in practice. Hence, detection of whether an image had undergone such filtering enhancement is important for forensic analysis and complements the content forgery detection techniques (e.g., those that detect copy-move forgery). Median filtering is the most popular enhancing method to hide the traces of splicing forgery, and therefore finding such filtering gained much research focus of late [11]. An example of median filtering and face splicing attack has been shown in Fig. 1.7.

#### 1.3.2.5 Retouching Image Forgery

Retouching implies enhancing or hiding specific image features, e.g., lighting condition, illumination, texture in order to make the image attractive. Easy availability of commercial or free software performing such operations has made these type of forgery very common, especially in social media applications. Example of such forgery is shown in Fig. 1.8, where the face in the picture has been retouched to make it more attractive.

#### 1.3.2.6 Resampling Image Forgery

Resizing some specific object in the image or the whole image with different height and width is done in resampling image forgery. Resampling however just changes

**Fig. 1.8** Retouching image forgery: original image (left) and retouched version (right)

**Fig. 1.9** Resampling image forgery: original image (left) and resampled version (right)

the dimension of the image, but do not improve or decrease the quality of the image. An example of resampling image forgery has been shown in Fig. 1.9.

#### 1.3.2.7 Computer-Generated Image

Today's advanced technology and software can generate images which look exactly like real (photographic) images and are hard to distinguish from photographic images visually. Forensic examination of images to verify whether they are computer-generated or not is therefore an important and challenging research problem. Examples of such computer-generated images are shown in Fig. 1.10 [3].

In the next chapter, we describe detailed background on the problem of camera source identification in detail and describe a novel machine learning-based technique of solving this problem through *Discrete Cosine Transform Residual* (DCTR) features.

**Fig. 1.10** Examples of computer-generated images

# References


1. Birajdar, G.K., Mankar, V.H.: Digital image forgery detection using passive techniques: a survey. Digit. Investig. **10**(3), 226–245 (2013)
2. Cox, I., Miller, M., Bloom, J., Fridrich, J., Kalker, T.: Digital Watermarking and Steganography, vol. 2. Morgan Kaufmann Publishers (2008)
3. Karras, T., Aila, T., Laine, S., Lehtinen, J.: Progressive growing of gans for improved quality, stability, and variation (2017). *arXiv preprint* arXiv:1710.10196
4. Metro Journalist: Morphed photo of Obama. https://ametrojournalist.wordpress.com/2014/02/06/morphed-photo-of-obama-obama-watching//-modis-speech-namo-metro-journalist/
5. Naskar, R., Malviya, P., Chakraborty, R.S.: Digital forensics: state-of-the-art and open problems. Innovative Research in Attention Modeling and Computer Vision Applications, pp. 260 (2015)
6. Roy, A., Chakraborty, R.S.: Optimal distortion estimation for prediction error expansion based reversible watermarking. In: International Workshop on Digital Watermarking, pp. 265–279. Springer (2016)
7. Roy, A., Chakraborty, R. S.: Towards optimal prediction error expansion based reversible image watermarking. In: IEEE Transactions on Circuits and Systems for Video Technology (2019). https://doi.org/10.1109/TCSVT.2019.2911042
8. Roy, A., Chakraborty, R.S., Naskar, R.: Reversible color image watermarking in the YCoCg-R color space. In: International Conference on Information Systems Security, pp. 480–498. Springer (2015)
9. Roy, A., Maiti, A.K., Ghosh, K.: A perception based color image adaptive watermarking scheme in YCbCr space. In: 2015 2nd International Conference on Signal Processing and Integrated Networks (SPIN), pp. 537–543. IEEE (2015)
10. Roy, A., Maiti, A.K., Ghosh, K.: An HVS inspired robust non-blind watermarking scheme in YCbCr color space. Int. J. Image Graph. **18**(03), 1850015 (2018)
11. Roy, A., Tariang, D.B., Chakraborty, R.S., Naskar, R.: Discrete cosine transform residual feature based filtering forgery and splicing detection in JPEG images. In: Proceedings of the IEEE Conference on Computer Vision and Pattern Recognition Workshops, pp. 1552–1560 (2018)

12. Tariang, D.B., Roy, A., Chakraborty, R.S., Naskar, R.: Automated JPEG forgery detection with correlation based localization. In: 2017 IEEE International Conference on Multimedia & Expo Workshops (ICMEW), pp. 226–231. IEEE (2017)
13. The New York Times: In an iranian image, a missile too many. https://thelede.blogs.nytimes.com/2008/07/10/in-an-iranian-image-a-missile-too-many/

# Chapter 2
# Camera Source Identification

## 2.1 Introduction

Camera source identification is a classic forensic problem [20]. Blind detection of camera source of an image is used to verify its origin and authenticity. Hence in criminal investigations, forensic analysis of evidences requires efficient camera source identification techniques. Therefore, such techniques are primarily used for forensic analysis in military and defense applications.

In the next section, we shall discuss the state-of-the-art digital forensic techniques for analysis of image origin and its source identification. First, we shall briefly discuss the various components of a present-day image capturing device, i.e., the digital camera.

## 2.2 Digital Camera Components

The two major components of an image capturing device are the lens and the sensor. Below we discuss their roles in image formation and storage, in a digital camera [1]:

- **Taking Lens**: It forms an image of the scene (in front of the camera) on the surface of the sensor. The taking lens is responsible for bending and curving the path of incident radiation, so as to make it reach the sensor.
- **Sensor**: It converts incident radiation into photocharges, which are subsequently digitized and stored as raw image data. Basic imaging element of a sensor is a "pixel" and a sensor is made up of a rectilinear grid of pixels. The sensor is responsible for sensing a sampled version of the image formed. A good sensor generates photocharges at precise locations in the image and also prevents subsequent random dispersion of these charges until they are read out. After capturing the incident radiation in the form of photocharges in a digital camera sensor, what remains is to acquire the color information about the image. This is done by color filters.



A. Roy et al., *Digital Image Forensics*, Studies in Computational Intelligence 755, https://doi.org/10.1007/978-981-10-7644-2_2

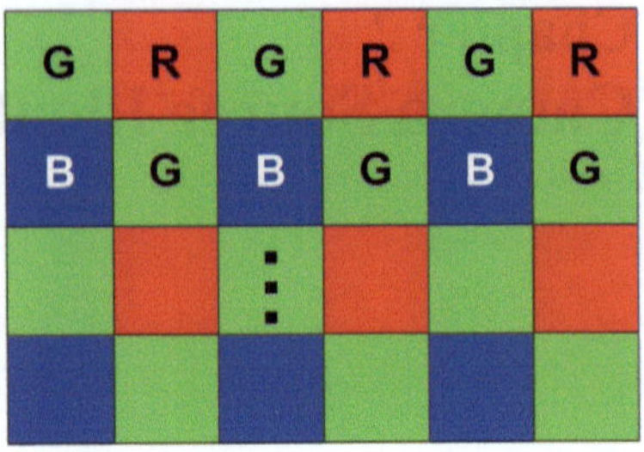


**Fig. 2.1** Bayer pattern arrangement of color filters

- **CFA**: Color filters are nothing but (single or multiple) thin layer(s) of different colorants, laid on the surface of the camera sensor. An array of pixel-sized color filters, arranged over the sensor surface, enables detection of full color information of an image, using only a single sensor. This array is called CFA or Color Filter Array.

Since a color image can be fully digitally represented by three components, for, e.g., RGB (Red Green Blue), precise capture of full color information of an image requires a set of at least three sensors at each pixel location. However, this specification would highly increase the cost and complexity of manufacturing the device. Therefore, in conventional digital cameras, each pixel location is designed to sample only one color. Hence, the cost and complexity are reduced by the use of single color filter at each pixel location; the missing colors are interpolated from the neighboring pixel locations. CFA interpolation is often termed as *demosaicing*. The Bayer pattern [3], shown in Fig. 2.1, is one such example of arrangement strategy for color filters over a sensor CFA. The Bayer pattern uses 50% green filters and 25% filters to sense each of red and blue color, taking into account the fact that the Human Visual System (HVS) is more sensitive to the green component of an RGB image. In this pattern, at each pixel location, one color is precisely recorded and other two colors are interpolated from the neighboring pixels.

After collection of color information of an image is over and the interpolated image is available, some further image enhancement steps [1], such as sensor noise reduction, color correction, edge enhancement, etc., are carried out before the finished image is finally produced in the camera.

Information about the image and its source may be obtained from image metadata, such as the file header or EXIF (Exchangeable Image File Format). Such metadata include information such as camera make and model (but not the specific camera identity), manufacturer details, image size, date and time of image capture, compression ratio in JPEG images, etc. However, such information contained in the file header is highly vulnerable to modification attacks. Hence, we need blind forensic techniques to throw some light on the origin of an image. Next, we shall present the state-of-the-art digital forensic techniques, proposed for image source investigation.

Blind forensic techniques exploit the different traces left behind in an image, as a result of various processing steps, while its formation and storage by a particular device. Those traces or artifacts are formed in an image in either one of two steps:

- **Image Acquisition**: Image acquisition artifacts are nothing but either the distortions generated by the camera lens while capturing a scene, or the sensor noise produced during color information acquisition (and interpolation) by the CFA.
- **Image Storage**: Image storage artifacts are those patterns which are generated in an image while processing it for storage in a standard format. Such processing activities include noise removal, image enhancement, etc.

The image artifacts described above, function as a means of distinguishing between different camera models, depending on their components such as the camera lens. Those artifacts may also provide specific device-related information, utilizing the traces produced by the sensor.

## 2.3 Literature Review

In the literature, there exist several methodologies for efficient detection of the camera source. These techniques try to detect traces or artifacts that the source camera imparts onto the images. Several techniques are proposed to identify the camera models by finding out such artifacts using models of the camera's demosaicing algorithms [4, 27], imaging sensor noise [10], lens-induced chromatic aberration [25, 30], and proprietary implementations of JPEG compression [19]. Kharrazi et al. [20] used several *Image Quality Metric* (IQM) and *Non-Image Quality Metric* (Non-IQM) features-based classification technique for source camera identification. Chen et al. [6] recently used SRMQ1 features utilizing the rich model of camera's demosaicing algorithm for efficient source camera identification, inspired by Fridrich et al. [13] work on rich models for steganalysis. Fridrich et al. [13] developed the SRMQ1 features utilizing the camera's *Color Filter Array* (CFA) pattern and color interpolation algorithm for steganalysis. Chen et al. [6] also found such features efficient for source camera identification. Co-occurrence matrix-based local features have been used by Marra et al. [22] for efficient camera model identification. Recently, in [28] Tuama et al. extracted higher order statistics from contaminated sensor noise by computing co-occurrence matrix-based features for source camera identification and further improved the performance by additionally incorporating color dependencies and conditional probability-based features in [29].

The general processing pipeline in a digital camera is shown in Fig. 2.2. Light reflected from the natural scene passes through the camera lens followed by Color

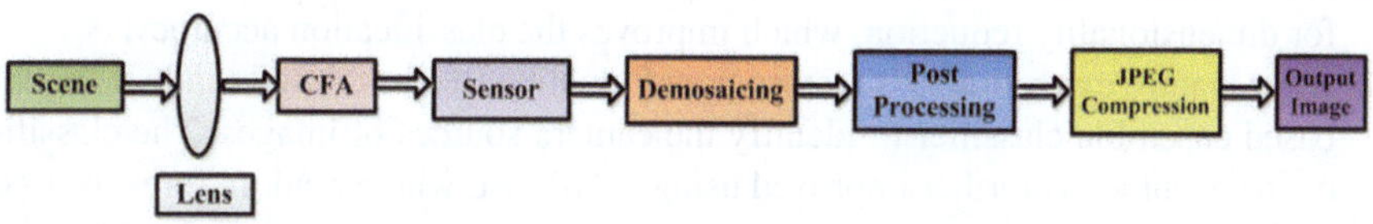


**Fig. 2.2** Typical digital camera image processing pipeline

Filter Array (CFA) and sensor. CFA allows only one color component of the light per pixel to pass through before reaching the sensor, and other two unobserved color values at each pixel positions are interpolated using a demosaicing process. That is why rich model using camera's demosaicing features traces camera intrinsic parameters, which is indeed relevant for camera source identification as verified by Chen et al. [6]. After the demosaicing and some post-processing steps, the image is stored in the camera in compressed format. Currently, JPEG is the most popular compression technique which uses *Discrete Cosine Transform* (DCT) based image compression with quantization [15]. Generally, the quantization tables used for the JPEG compression vary in different camera models. Therefore, JPEG compression incurs camera model dependent quantization noise in the image also. Consequently, these processing steps incur camera dependent artifacts in the image.

The goal of the forensic techniques is to identify traces of the artifacts and hence to identify the source camera model. Generally, the techniques in the literature develop a machine learning-based parametric model of specific components or artifacts. Next, the parameters are selected for each image by training a large dataset with different camera models taken as levels. Finally, the parameter values are used as features to train a classifier for identifying the image camera source.

Recently, Fridrich et al. [17] proposed a low-complexity *Discrete Cosine Transform Residual* (DCTR) feature utilizing first order statistics of quantization noise residual obtained from the decompressed JPEG image, using 64 kernels of the discrete cosine transform for JPEG image steganalysis. As these features use the first order statistics of quantization noise, which is also an intrinsic parameter of different camera models, therefore, using these features result in highly accurate camera source classification. Moreover, since the dimension of the DCTR feature space is relatively high (e.g., feature vector with 8000 components), incorporation of *Principal Component Analysis* (PCA) based feature transformation for dimensionality reduction additionally improves the classification accuracy.

In this chapter, DCTR features are used along with PCA-based dimensionality reduction has been utilized for efficient camera source identification. Moreover, after feature extraction and feature transformation using PCA, random forest-based ensemble classifier along with *Adaptive Boosting* (AdaBoost) has been used to predict the camera sources.

The major contributions of the chapter can be summarized as follows:

- In this chapter, DCTR features, originally proposed by Fridrich et al. [17] for steganalysis, have been used for our source camera identification problem which outperforms the state-of-the-art approach [6].
- After feature extraction, we have incorporated PCA-based feature transformation for dimensionality reduction, which improves the classification accuracy.
- After feature extraction and feature transformation, we have used random forest-based ensemble classifier to identify the camera sources of images. The classification accuracy is further improved using AdaBoost, where random forest is used as the base learner.

The rest of the chapter is organized as follows. In Sect. 2.4, the DCTR feature-based source camera identification framework is discussed in detail. Experimental results and comparison with the state of the art have been given in Sect. 2.5, and finally, the chapter concludes in Sect. 2.6.

## 2.4 Source Camera Identification Framework

In this section, we describe the source camera identification framework [24] in detail.

### *2.4.1 Motivation for Choice of Features*

Figure 2.2 shows the processing pipeline in a typical modern digital camera. Reflected light from a natural scene passes through the lens to reach the sensor via a *Color Filter Array* (CFA). Next, after demosaicing and post-processing, the image is saved in a compressed format. JPEG being the most popular compression format, we are concerned with the JPEG images here. JPEG compression [15] basically divides the whole image into $8 \times 8$ blocks and applies *Discrete Cosine Transform* (DCT) into each block. Subsequently, each block is quantized according to a given quantization table, which varies for different camera models [18]. As a result, the camera model specific quantization table incurs artifacts as quantization noise in the JPEG compressed image. Fridrich et al. [17] used this idea to propose a new low-complexity feature set named *Discrete Cosine Transform Residual* (DCTR) for steganalysis. The DCTR feature is basically the first order statistics of quantization noise residuals obtained from the decompressed JPEG images using 64 kernels of the discrete cosine transform. Hence, these features effectively capture the traces of quantization noise incurred by the camera source. This is the primary motivation for choosing DCTR feature for source camera identification problem and through experimental results, we verify its efficacy.

### *2.4.2 DCTR Feature Extraction*

Now we shall discuss DCTR feature extraction in detail in this section. To generate the DCTR features, the original image in spatial domain is convolved with the 64 DCT basis patterns, each of size $8 \times 8$, to get 64 new undecimated DCT planes, i.e., DCT residuals. The final features are generated as histograms of the DCT residuals. The steps for DCTR feature extraction from an image is given below [17, 26]:

- The JPEG image is decompressed to spatial domain without quantizing the pixel values to $\{0, \ldots, 255\}$ to avoid any loss of information.

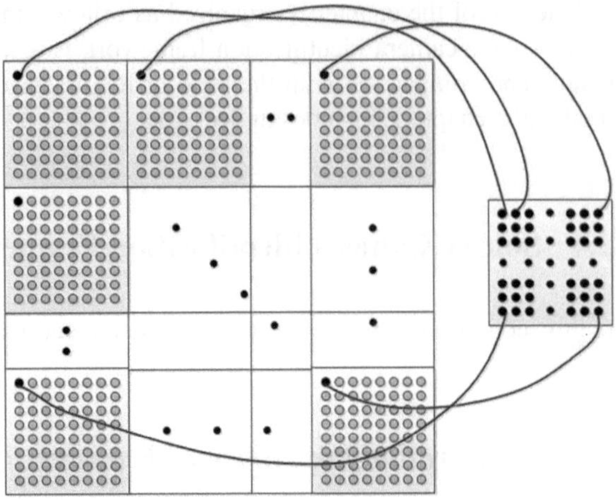

**Fig. 2.3** Subsampling procedure

- The DCT basis patterns of size 8 × 8 are generated as $\mathbf{B}^{(k,l)} = \left(B_{mn}^{(k,l)}\right), 0 \leq m, n \leq 7$:

$$B_{mn}^{(k,l)} = \frac{w_k w_l}{4} \cos \frac{\pi k (2m+1)}{16} \cos \frac{\pi l (2n+1)}{16} \tag{2.1}$$

where $w_0 = \frac{1}{\sqrt{2}}$, $w_i = 1$ for $i > 0$.
- The decompressed JPEG image $\mathbf{X}$ is convolved with each of the 64 DCT basis patterns $\mathbf{B}^{(k,l)}$, to generate a set of 64 undecimated DCT, each of which is denoted by $\mathbf{U}^{(k,l)}$ for $(k,l)$-th DCT basis pattern as

$$\mathbf{U}^{(k,l)} = \mathbf{X} * \mathbf{B}^{(k,l)}, \qquad 0 \leq k, l \leq 7 \tag{2.2}$$

- According to the 64 DCT modes $(a, b), 0 \leq a, b \leq 7$, corresponding to each DCT basis pattern in each 8 × 8 DCT block, the filtered undecimated DCT image $\mathbf{U}^{(k,l)}$ is subsampled by a step-size of 8 to get 64 sub-images $\mathbf{U}_{a,b}^{(k,l)}$, as shown in Fig. 2.3.
- For each sub-image $\mathbf{U}_{(a,b)}^{(k,l)}$, the histogram feature is extracted as

$$\mathbf{h}_{a,b}^{(k,l)}(x) = \frac{1}{|\mathbf{U}_{a,b}^{(k,l)}|} \sum_{u \in \mathbf{U}_{a,b}^{(k,l)}} [Q_T(|u|/q) = x] \tag{2.3}$$

where $Q_T$ is a quantizer with integer centroids $\{0, 1, \ldots, T\}$, $q$ denotes the quantization step, and $[P]$ is the *Iverson bracket*, which is equal to "0" when the statement $P$ is false, and "1" when $P$ is true. Here, $q$ is dependent on the JPEG quality factor also [17].
- All the histogram features of the 64 sub-images $\mathbf{U}_{a,b}^{(k,l)}$ are merged and combined to obtain the histogram feature $\mathbf{h}^{(k,l)}$ of the filtered undecimated DCT image $\mathbf{U}^{(k,l)}$.

- The histogram features extracted from the 64 sub-images can be merged [17] to reduce the feature dimension using symmetry properties. Authors in [17] used DCTR feature dimension to be 8000 as a good compromise between performance and detectability for efficient steganalysis.

The primary advantage of DCTR feature is its low complexity and better detection accuracy with less dimension w.r.t all JPEG steganography algorithms [17]. This feature has been used to identify source camera models of images as this feature involves first order statistics of quantized noise residuals.

### 2.4.3 Feature Transformation by PCA

Dimensionality reduction generally transforms high-dimensional feature space to lower dimensional feature space to avoid overfitting and "curse of dimensionality". *Principal Component Analysis* (PCA) is a linear method of feature transformation as well as dimensionality reduction, which maps a linear transformation from the higher dimensional feature space to the lower dimensional feature space, such that the variance of the lower dimensional feature space is maximized [9]. Generally, the covariance matrix of the higher dimensional feature space is obtained, and the eigenvectors of the covariance matrix are calculated. Then, the eigenvectors corresponding to the largest eigenvalues are selected as principal components. In this way, dimensionality is reduced which in turns reduce time and space overhead. Moreover, removal of correlation among feature space improves the performance of the machine learning models.

The DCTR feature, consisting of a relatively high-dimensional feature space (8000 components in the feature vector), can suffer from the problem of overfitting, while machine learning model building due to the "curse of dimensionality" [9]. Therefore, we have incorporated PCA-based feature transformation and dimensionality reduction to make the data points decorrelated by maximizing the variance among them and efficient machine learning model building. Selecting the number of components in the low-dimensional feature space ("principal components") is an important optimization decision. For that purpose, we have used the *"95% variance method"* [2], i.e., if the $m$-dimensional sample data points are $x^{(i)}$, and corresponding projected data points are denoted by $x^{(i)}_{approx}$, then we iterate the process until:

$$\frac{\frac{1}{m}\sum_{i=1}^{m}||x^{(i)} - x^{(i)}_{approx}||^2}{\frac{1}{m}\sum_{i=1}^{m}||x^{(i)}||^2} \leq 0.05 \tag{2.4}$$

A computationally easier method incorporates singular value decomposition of the covariance matrix of the data points, and finding the largest $k$ number of singular values that satisfy Eq. (2.5) given below, which is equivalent to Eq. (2.4) [2]:

$$\frac{\sum_{i=1}^{k} s_{ii}}{\sum_{i=1}^{m} s_{ii}} \geq 0.95 \tag{2.5}$$

Here, $k$ denotes the number of principal components selected automatically such that among the data points 95% variance is retained, and $S = [s_{ij}]$ denotes the singular value matrix after singular value decomposition of the covariance matrix of the feature set. We later show that PCA significantly improves the classification accuracy of our experiments by 4–5%.

## 2.4.4 *Classification*

Camera source identification is actually a classification problem, where the correct class of camera source model has to be predicted. A brief description of the classification process and the effective classifiers has been provided below.

### 2.4.4.1 K-Fold Cross-Validation

To overcome the problem of overfitting, "$K$-fold cross-validation" is applied to the dataset. "$K$-fold cross-validation" basically means the dataset is divided approximately equally into $K$ parts. Then $(K - 1)$ parts are used for training and the remaining part is used for testing. This is repeated for the $K$ different parts individually, and the average accuracy is reported. In our case, we have used 10-fold cross-validation.

### 2.4.4.2 Random Forest Classifier

After feature extraction, we have used *random forest*-based multi-class ensemble classifier [7]. For steganalysis, Kodovsky et al. [21] used multi-class ensemble classifier by grouping set of binary ensemble classifier together and performing majority voting among the outputs of the base learners, in conjunction with *Fisher Linear Discriminant* (FLD) classifier as base learners. We used random forest-based multi-class ensemble classifier with decision tree as base learner. Random forest classifier has several advantages [7], viz.,

- They can represent nonlinear decision boundaries.
- Those are efficient in computation and memory usages during training and testing.

- They perform integrated feature selection and classification.
- They are resilient in the presence of noisy features.

In the training phase, random forest classifier selects random input sample features from the training set and builds decision trees. Information gain (entropy based) determines which attribute in a given set of training feature vectors is most useful for discriminating between classes to be learned. Then the attribute with highest information gain is chosen to be the root of the decision tree. The child nodes are constructed as subsets of vectors consisting of each value of the root attribute. In this way, decision trees are constructed as trained classifiers.

In the testing phase, test data is fed to each of the decision trees of the trained classifier, and a predicted class is obtained from each decision tree in the ensemble classifier. Then, the final output class is predicted taking an average or aggregation of the predictions from the individual trees of the ensemble.

### 2.4.5 Ensemble Classifier

Instead of using a single classifier, if we merge the decisions of multiple classifiers, then it can potentially improve classification accuracy. Hence, ensemble classifiers are used in practice. There are mainly two well-known techniques for ensemble classification. Those are (a) bagging and (b) boosting.

#### 2.4.5.1 Bagging

*Bagging* is also known as *bootstrap aggregating* [23]. In this technique, the original training data is the first bootstrap sampled, then each bootstrap sample is used to fit a separate base classifier, and finally the base classifiers are combined by taking majority voting or aggregation of the predictions from each base classifier. Each bootstrap sample is generated by randomly sampling the training data with replacement. Bagging reduces variance and also avoids overfitting. *Random Forest* is an extension of bagging [23], where the decision tree is taken as base classifier, and the prediction of the ensemble classifier is taken as the majority voting among the base classifiers. Additionally, feature selection is carried out at each node while constructing the decision tree. The random forest classifier generally does not overfit if the number of trees increases.

#### 2.4.5.2 Boosting

*Boosting* is a weighted ensemble classifier, where each base classifier is assigned a nonnegative weight according to the importance of the base classifier [11]. This approach generates a strong classifier by combining several weak base classifiers. The

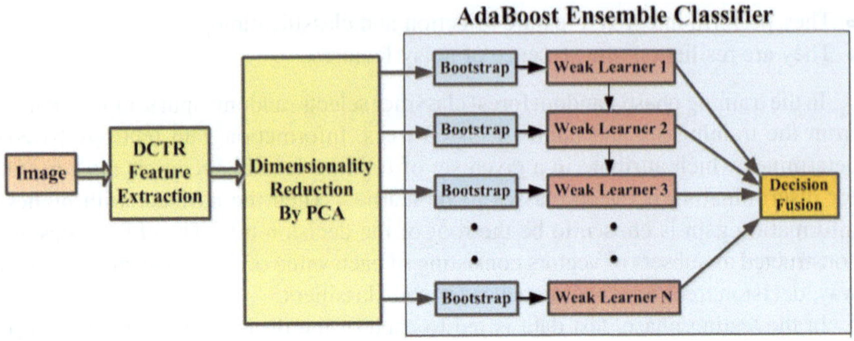


**Fig. 2.4** Flowchart of ensemble classifier training process

most popular representation of boosting is *AdaBoost* [11]. Classically, AdaBoost was designed for binary classification problem. However, it is generalized for multi-class classification as *AdaBoost.M1* [11].

AdaBoost is the ensemble of weak learners. Weak learning implies it performs only slightly better than a randomly guessing classifier (i.e., 50% accuracy). This algorithm in each iteration learns hypotheses from these weak learners. The final hypothesis is a weighted vote of the weak hypotheses, where the weights of the hypothesis are computed as a function of the errors corresponding to each hypothesis [11]. In this way, combining the weak learners, a strong classifier is constructed with much better accuracy. We have used the AdaBoost ensemble classifier with Random Forest as weak learner for training in our source identification framework as shown in Fig. 2.4. After feature extraction and transformation by PCA, each bootstrap sample in the feature space is used to fit random forest-based weak learner, then ensemble classifier is obtained by AdaBoost.

## 2.5 Experimental Results

### *2.5.1 Experimental Setup*

In this work, we have experimented with 10,507 images taken from the standard *Dresden Image Database* [14] (termed as **dataset-1**) for 10 camera models as shown in Table 2.1. In order to resist overfitting and sufficient data for the machine learning-based model building, around 630–1671 images were taken from each camera. Further, we extend our experimentation to effectively identify the camera brand attribute of the images. The colored JPEG images were transformed into gray scale images prior to DCTR feature extraction with quality factor has been set to $Q = 75$. Several works lacked the generalized comparison among the effectiveness of the features for this particular problem by experimenting with their own individual image dataset

**Table 2.1** Camera models used in experiments

| Camera ID | Number of image | Camera model |
|---|---|---|
| 1 | 979 | Canon |
| 2 | 1040 | Olympus |
| 3 | 1331 | Samsung |
| 4 | 1671 | Sony |
| 5 | 1000 | Agfa |
| 6 | 925 | Casio |
| 7 | 630 | FujiFilm |
| 8 | 1000 | Kodak |
| 9 | 1000 | Nikon |
| 10 | 931 | Panasonic |

which is not publicly available [6]. Therefore, for fair comparison with other state-of-the-art schemes, we evaluated our approach using a reduced subset of 100 images from each of the 10 camera models from the Dresden Image Database (termed as **dataset-2**). It is noteworthy that dataset-1 is experimented for evaluating camera brand attribute identification (i.e., in general, camera brand identification, e.g., Nikon vs. Canon) whereas dataset-2 is utilized for efficient camera model attribute identification (i.e., specific camera model identification, e.g., Nikon-1 vs. Nikon-2). Naturally, dataset-2 would be a subset of dataset-1 with more variety of camera model fingerprints.

Experimental validation suggests that the proposed approach is efficient for both the camera brand attribute identification as well as particular camera model attribute identification. In the source identification framework, we have trained the Random Forest (RF) classifier (decision tree being the base learner) with the DCTR features and labels (camera model it originally belongs to) given for images in the training set. The random forest classifier is taken as an ensemble of 100 decision tree in our experiments. For AdaBoost, the learning rate is 1 and the number of weak learners is 50 , i.e., the ensemble learns at the maximum possible rate.

The dataset is split evenly into approximately 10 parts randomly, among them 9 parts are used for training and the remaining part is used for testing. This process is iterated for the 10 different possibilities individually, and finally, the average accuracy is reported (i.e., 10-fold cross-validation). The classification accuracy for the popular *Support Vector Machine* (SVM) classification, with *Radial Basis Function* (RBF) kernel, has also been evaluated.

In the experiments, MATLAB (v. R2015a) has been used, and the experiments were run on an iMac workstation with a 3.2 GHz Intel CPU and 8 GB of RAM. The code for the DCTR feature extraction was adapted from [12]. For the machine learning-based model building and experimental validation, the *Weka Machine Learning Toolbox* [16] has been used.

**Table 2.2** Comparison of accuracy for various classification schemes based on DCTR features (dataset-1)

| Experiment | Classification accuracy (%) |
|---|---|
| SVM classifier (RBF kernel) | 81.1 |
| AdaBoost (100 decision trees as weak learner) | 45.2 |
| Random Forest (RF) (100 decision trees as base learner) | 93.8 |
| RF + PCA (168 dim. features) | 98.9 |
| AdaBoost (RF as weak learner) + PCA (168 dim. features) | 99.1 |

### 2.5.2 *Classification Accuracy Improvement (Dataset-1)*

The classification performance for camera source identification on dataset-1 is compared rigorously in Table 2.2. We evaluate the following classification accuracy results for the DCTR feature sets extracted (we explicitly mention whether dimensionality reduction was used):

- Without dimensionality reduction, SVM classification (with RBF kernel) results an accuracy of 81.1%.
- AdaBoost ensemble classifier (with 100 decision tree as weak learner) without dimensionality reduction provides a classification accuracy of only 45.2%.
- Random Forest (RF) classification (decision tree being base learner) without dimensionality reduction results an average classification accuracy of 93.8%.
- Random forest classifier, in conjunction with dimensionality reduction through Principal Component Analysis (PCA), enhances the average classification accuracy up to 98.9%. The automated optimal dimensionality was achieved (i.e., 168 in this case) by the "Modified 95% Variance" method, as depicted in Fig. 2.5.
- Finally, if we employ *AdaBoost* with random forest as weak learner, combining dimensionality reduction through PCA, the classification accuracy further enhances to **99.1%**. The corresponding confusion matrix for the 10-class camera source classification problem has been illustrated in Table 2.3.

### 2.5.3 *Comparison of Accuracy with State-of-the-Art Techniques (Dataset-2)*

The proposed scheme has been compared and evaluated with the state-of-the-art approach reported in [6] using co-occurrence-based *Spatial and Color Rich Model with Fixed Quantization* $q = 1$ (SRMQ1) for dataset-2. Both the co-occurrence-based features and the DCTR features were extracted from the images of dataset-2

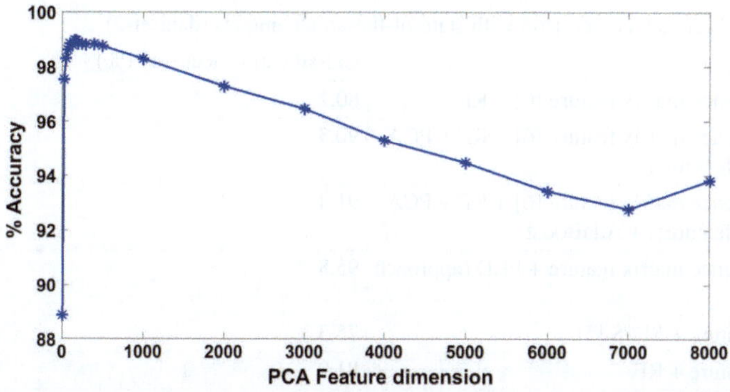


**Fig. 2.5** Classification accuracy versus PCA dimension plot. Accuracy is maximized when PCA dimension is 168

**Table 2.3** Confusion matrix for the proposed classification scheme

| Classification accuracy | | True model | | | | | | | | | |
|---|---|---|---|---|---|---|---|---|---|---|---|
| | | 1 | 2 | 3 | 4 | 5 | 6 | 7 | 8 | 9 | 10 |
| Identified model | 1 | **98.8** | 0 | 0.2 | 0 | 0.6 | 0.1 | 0 | 0 | 0.3 | 0 |
| | 2 | 0 | **99.8** | 0 | 0 | 0.1 | 0 | 0 | 0 | 0.1 | 0 |
| | 3 | 0.2 | 0.1 | **98.9** | 0 | 0.5 | 0 | 0 | 0 | 0.2 | 0.1 |
| | 4 | 0 | 0 | 0 | **99.8** | 0.2 | 0 | 0 | 0 | 0 | 0 |
| | 5 | 0.3 | 0 | 0.8 | 0.4 | **96.9** | 0.4 | 0 | 0.1 | 1.0 | 0.1 |
| | 6 | 0.1 | 0 | 0 | 0.1 | 0.1 | **99.7** | 0 | 0 | 0 | 0 |
| | 7 | 0.2 | 0 | 0 | 0 | 0.1 | 0 | **99.5** | 0 | 0.2 | 0 |
| | 8 | 0.1 | 0 | 0 | 0 | 0.1 | 0 | 0 | **99.8** | 0 | 0 |
| | 9 | 0.2 | 0 | 0 | 0 | 1.6 | 0 | 0 | 0.1 | **98.1** | 0 |
| | 10 | 0 | 0 | 0.1 | 0 | 0.2 | 0 | 0 | 0 | 0.2 | **99.5** |

and used to train the same RF classifier (with the number of trees being 100) for fair and effective comparison. The overall resulting classification accuracies have been summarized in Table 2.4, which clearly implies that the proposed technique is more effective than the co-occurrence matrix-based feature approach [6], with respect to the same RF-based classifier. We have also experimented with different combinations of the choice of classifier as well. The proposed DCTR feature-based approach (along with dimensionality reduction through PCA and AdaBoost) outperforms the multi-class ensemble classifier using *Fisher Linear Discriminant* (FLD) based binary classifiers proposed in [6]. Also, the proposed technique is found to be quite effective identifying particular camera models (particular camera model identification is considered for dataset-2).

Several popular dimensionality reduction techniques have been exploited to obtain the most suitable alternative along with the choices of classifiers for classification

**Table 2.4** Accuracy comparison with state-of-the-art techniques (dataset-2)

| Approach | Classification accuracy (%) |
|---|---|
| Co-occurrence matrix feature [6] + RF | 80.1 |
| Co-occurrence matrix feature [6] + RF + PCA (100 dim. features) | 90.3 |
| Co-occurrence matrix feature [6] + RF + PCA (100 dim. features) + AdaBoost | 91.4 |
| Co-occurrence matrix feature + FLD (approach of [6]) | 95.8 |
| DCTR Feature + SFFS [5] | 75.3 |
| DCTR Feature + RF | 81.9 |
| DCTR Feature + RF + AdaBoost | 83.4 |
| DCTR Feature + FLD | 95.6 |
| DCTR Feature + PCA (130 dim. features) + FLD | 96.2 |
| DCTR Feature + RF + PCA (130 dim. features) | 96.3 |
| DCTR Feature + RF + PCA (130 dim. features) + AdaBoost | 96.5 |

accuracy improvement. Table 2.4 clearly suggests that elementary techniques, e.g., the *Subset Size Feature Selection* (SFFS) technique, are not quite useful in this regard for feature selection. The reason being the high feature dimensionality of DCTR (8,000 dimension feature) which rendered the overall scheme inefficient. However, dimensionality reduction through PCA (reduces feature dimension from 8000 to 130 for dataset-2) significantly enhances the classification accuracy as indicated in Table 2.4, especially when AdaBoost is applied along with.

### 2.5.4 Evaluation of Overfitting Trends

We additionally check for the robustness of the model to overfitting. For the sake of that, we took the classifier (DCTR + PCA (168) + RF (number of trees = 100) + AdaBoost) built using dataset-1, and then evaluated the model for **images from a completely different *Sensor Dust Image Dataset*** [8], for two camera models: *Kodak-M1063* and *Canon-Ixus55*. The average classification accuracy obtained was 97.08%, which signifies that the classification model built is not prone to overfitting and has good generalization capability.

## 2.6 Conclusion

Camera source identification is an important forensic problem, in which the blind investigation of camera source or the class of camera source has to be made from the image under question. In this chapter, we have developed a DCTR feature-based source camera identification framework, with random forest-based ensemble classification fortified by AdaBoost. The efficacy of the DCTR feature stems from the fact that different camera sources induce distinguishable JPEG compression artifacts on the image due to varying quantization table used for JPEG compression by the camera source. Moreover, since DCTR is a high-dimensional (i.e., 8000) feature set, dimensionality reduction of the features through PCA prior to classification model building improves the classification accuracy. The proposed technique outperforms other state-of-the-art techniques when applied to a large benchmark image dataset from 10 different camera sources from the *Dresden Image Dataset*, and also demonstrates low overfitting trends.

In the next chapter, we give a detailed description of the problem of copy-move forgery, along with three metrics to estimate the accuracy of a detection technique.

## References


1. Adams, J., Parulski, K., Spaulding, K.: Color processing in digital cameras. IEEE Micro **18**(6), 20–30 (1998)
2. Andrew N.G.: Machine learning. https://www.coursera.org/learn/machine-learning
3. Bayer, B.E.: Color imaging array. US Patent No. 3,971,065 (1976)
4. Cao, H., Kot, A.C.: Accurate detection of demosaicing regularity for digital image forensics. IEEE Trans. Inf. Forensics Secur. **4**(4), 899–910 (2009)
5. Çeliktutan, O., Sankur, B., Avcibas, I.: Blind identification of source cell-phone model. IEEE Trans. Inf. Forensics Secur. **3**(3), 553–566 (2008)
6. Chen, C., Stamm, M.C.: Camera model identification framework using an ensemble of demosaicing features. In: 2015 IEEE International Workshop on Information Forensics and Security (WIFS), pp. 1–6. IEEE (2015)
7. Criminisi, A., Shotton, J., Konukoglu, E.: Decision forests for classification, regression, density estimation, manifold learning and semi-supervised learning. Technical Report MSRTR-2011-114, Microsoft Research Cambridge, **5**(6), 12 (2011)
8. Dirik, A.E., Sencar, H.T., Memon, N.: Sensor dust image dataset. http://isis.poly.edu/~emir/sensordust/sensordust3.html
9. Duda, R.O., Hart, P.E., Stork, D.G.: Pattern Classification. Wiley (2012)
10. Filler, T., Fridrich, J., Goljan, M.: Using sensor pattern noise for camera model identification. In: 15th IEEE International Conference on Image Processing, pp. 1296–1299. IEEE (2008)
11. Freund, Y., Schapire, R.E., et al.: Experiments with a new boosting algorithm. ICML **96**, 148–156 (1996)
12. Fridrich, J.: Feature extractors for steganalysis. http://dde.binghamton.edu/download/feature_extractors/
13. Fridrich, J., Kodovsky, J.: Rich models for steganalysis of digital images. IEEE Trans. Inf. Forensics Secur. **7**(3), 868–882 (2012)
14. Gloe, T., Böhme, R.: The dresden image database for benchmarking digital image forensics. J. Digit. Forensic Pract. **3**(2–4), 150–159 (2010)

15. Gonzalez, R.C., Woods, R.E.: Digital Image Processing, 3rd edn. Prentice Hall (2008)
16. Hall, M., Frank, E., Holmes, G., Pfahringer, B., Reutemann, P., Witten, I.H.: The weka data mining software: an update. ACM SIGKDD Explor. Newsl. **11**(1), 10–18 (2009)
17. Holub, V., Fridrich, J.: Low-complexity features for JPEG steganalysis using undecimated DCT. IEEE Trans. Inf. Forensics Secur. **10**(2), 219–228 (2015)
18. ImpulseAdventure: JPEG compression quality from quantization tables. http://www.impulseadventure.com/photo/jpeg-quantization.html
19. Kee, E., Johnson, M.K., Farid, H.: Digital image authentication from JPEG headers. IEEE Trans. Inf. Forensics Secur. **6**(3), 1066–1075 (2011)
20. Kharrazi, M., Sencar, H.T., Memon, N.: Blind source camera identification. In: International Conference on Image Processing, 2004. ICIP'04. 2004, vol. 1, pp. 709–712. IEEE (2004)
21. Kodovsky, J., Fridrich, J., Holub, V.: Ensemble classifiers for steganalysis of digital media. IEEE Trans. Inf. Forensics Secur. **7**(2), 432–444 (2012)
22. Marra, F., Poggi, G., Sansone, C., Verdoliva, L.: A study of co-occurrence based local features for camera model identification. Multimedia Tools and Applications, pp. 1–17 (2016)
23. Okun, O.: Feature selection and ensemble methods for bioinformatics: algorithmic classification and implementations. Medical Information Science Reference (2011)
24. Roy, A., Chakraborty, R.S., Sameer, U., Naskar, R.: Camera source identification using discrete cosine transform residue features and ensemble classifier. In: 2017 IEEE Conference on Computer Vision and Pattern Recognition Workshops (CVPRW), pp. 1848–1854. IEEE (2017)
25. San Choi, K., Lam, E.Y., Wong, K.K.Y.: Source camera identification using footprints from lens aberration. In: Electronic Imaging 2006, pp. 60690J–60690J. International Society for Optics and Photonics (2006)
26. Song, X., Liu, F., Yang, C., Luo, X., Zhang, Y.: Steganalysis of adaptive JPEG steganography using 2D Gabor filters. In: Proceedings of the 3rd ACM Workshop on Information Hiding and Multimedia Security, pp. 15–23. ACM (2015)
27. Swaminathan, A., Wu, M., Liu, K.J.R.: Nonintrusive component forensics of visual sensors using output images. IEEE Trans. Inf. Forensics Secur. **2**(1), 91–106 (2007)
28. Tuama, A., Comby, F., Chaumont, M.: Source camera model identification using features from contaminated sensor noise. In: International Workshop on Digital Watermarking, pp. 83–93. Springer (2015)
29. Tuama, A., Comby, F., Chaumont, M.: Camera model identification based machine learning approach with high order statistics features. In: European Signal Processing Conference, pp. 1183–1187 (2016)
30. Van, L.T., Emmanuel, S., Kankanhalli, M.S.: Identifying source cell phone using chromatic aberration. In: IEEE International Conference on Multimedia and Expo, pp. 883–886. IEEE (2007)

# Chapter 3
# Copy-Move Forgery Detection in Digital Images—Survey and Accuracy Estimation Metrics

## 3.1 Introduction

Image forgery detection is another major challenge in digital image forensics. *Image splicing* [3, 22], *image retouching* [18], and *region duplication* or *copy-move forgery* [2, 6–8, 15, 16, 28, 30] is the most common types of image forgery. Image splicing is also known as a copy-paste method, in which an image is manipulated by using components of other images. In this type, changes made to original images can be detected by using statistics of images. Image retouching is all about enhancing the visual appearance by changing the features of a digital image. In case of copy-move image forgery, the parts of an image are copied over itself to hide some object of the image. Identifying such forgery is nontrivial because significant changes have not been made to image features. So we cannot identify the alteration by analyzing the image statistics because they will not be changed. After the tampering, there will be no significant change in the original characteristics because the texture, noise, and color components will be almost similar to original image.

For example, Fig. 3.1 shows a forest scene in which lions are moving in a group. A copy-move forgery has been applied to Fig. 3.1a, and the output forged image is shown in Fig. 3.1b, where the image of a lion has been copied from the front leftmost position of original scene and pasted onto it.

In this chapter, we analyze the existing region duplication or copy-move forgery detection methods for digital images. The operating principles of most of the existing algorithms are "block-based", i.e., based on the identification of identical region. Hence, here, we present the readers a detailed survey of "block-based" copy-move forgery detection methods for digital images, along with the comparison of their performance, through evaluation of a set of three metrics [4].

The remainder of this chapter is organized as follows. In Sect. 3.2, we present an overview of related literature. In Sect. 3.3, we discuss block-based copy-move forgery detection techniques and also classify them. In Sect. 3.4, we present three metrics for performance estimation of existing copy-move forgery detection methods.



A. Roy et al., *Digital Image Forensics*, Studies in Computational Intelligence 755, https://doi.org/10.1007/978-981-10-7644-2_3

(a) Original image (b) Copy-move attack (duplicated object highlighted)

**Fig. 3.1** Examples of region duplication attack

In Sect. 3.5, we discuss the performance evaluation of different block-based region duplication detection algorithms with the help of these metrics. Finally, we conclude in Sect. 3.6.

## 3.2 Overview of Existing Techniques

Recently, there has been an extensive rise in the field of digital image forensics [5, 17, 19]. Initially, Fridrich et al. [6] proposed four different methods, viz., exact match, autocorrelation, exhaustive search, and robust match (based on Discrete Cosine Transform (DCT)) to find the copy-move forged regions in digital images. The robust match method has been proven to be better than the others because it allows finding the duplicate regions more accurately. The drawback of this method is that when we apply this method on large identical texture images it may give a large number of false matches. Popescu and Farid [16] developed a forgery detection method based on Principal Component Analysis (PCA). This method does not perform well for lossy compression because of the dimensionality reduction feature of PCA. Kang and Wei [8] developed a copy-move forgery detection method utilizing Singular Value Decomposition (SVD) features of digital image. This method is capable to detect forgery even image having slightly noise with low computational complexity. Zhang et al. [30] applied Discrete Wavelet Transform (DWT) on a forged image by decomposing it into four frequency sub-bands, and further divided approximate sub-band into overlapping blocks. Copy-move regions were detected using correlation values among overlapped block. The computational complexity of this technique is low as compared to the other existing algorithms.

Another notable wavelet-based methodology has been developed by Yang et al. [28], who decomposed a given test image into four sub-bands using Dyadic Wavelet Transform (DyWT). The Zernike coefficients of approximate sub-band are considered as the features of the test image. The features are then sorted lexicographically, and similarity between block pairs is calculated using the Euclidean distance. This algorithm gives less number of false positives, even when image has texture pattern. Bayram et al. [2] proposed a region duplication identification method based on Fourier Mellin Transform (FMT), which is invariant to rotation, translation, and scaling. However, this method can detect duplicate blocks with rotational angle up to 10 and scale factor up to 10%.

Ling et al. [12] developed a copy-move forgery detection algorithm, based on the principle of principal component analysis. This method gives satisfactory results for noisy and compressed images with low computational complexity. In the paper [7], Huang et al. have developed a region duplication detection algorithm that can detect forgery in digital image having additive white Gaussian noise [11] even distorted by JPEG compression [24].

## 3.3 Classification of Block-Based Copy-Move Forgery Detection Techniques

In this section, we classify existing block-based copy-move forgery detection algorithms on the basis of their operating principles, and then we describe the basic steps of block-based copy-move forgery detection algorithms selected from each category. The algorithms which are related to block-based copy-move forgery detection techniques may be divided into three categories as follows:

1. **Dimensionality Reduction-based Algorithms**. The operating principle of this category of region duplication forgery detection techniques is based on the reduction of dimensionality of image features, without losing meaningful information. The principle of dimensionality reduction helps to attain low computational complexity of finding copy-move regions. The existing forgery detection methods related to this category include the principal component analysis-based method [16], the singular value decomposition-based method [8], and the principal component analysis—discrete cosine transform-based method proposed [21].
2. **Discrete Cosine Transform (DCT) based Algorithms**. The algorithms related to this category are based on the operating principle of discrete cosine transform. Initially, the gray scale input image is divided into overlapping blocks and features are extracted (in the form of quantized DCT coefficients) from each block. The quantized coefficients contain meaningful information and play an important role to detect copy-move region. Examples of these DCT-based algorithms are [6], improved DCT based algorithms, and efficient DCT-based methods [25].
3. **Wavelet Transform-based Techniques**. In this category of copy-move forgery detection methods, initially, the forged image is decomposed into sub-bands using *wavelet transform*. Among these sub-bands approximate sub-band plays the major role to find copy-move forgery detection because of low-frequency content. Dyadic wavelet transform-based method [15], dyadic wavelet transform with Zernike moment-based method [28] and Complex Wavelet Transform (CWT) based method [27] are representatives of this class.

### *3.3.1 General Processing Pipeline for Copy-Move Forgery Detection Techniques*

Although the block-based region duplication detection techniques can be classified into different types, still they follow a generic processing pipeline which as described below:

1. Let us consider a forged image as an input image of size $W \times H$ pixels, if colored then convert into gray scale image, using the formula [14]:

$$I = 0.299 \times \text{red intensity} + 0.587 \times \text{green intensity} + 0.114 \times \text{blue intensity} \quad (3.1)$$

2. The gray scale image is divided into fixed size overlapping blocks.
3. The features of each block are extracted using the particular category to which the concerned detection method belongs, such as PCA, SVD, DCT, DyWT, Zernike moment, and CWT.
4. In the next step, all features are sorted using any one of the different methods, such as ascending, descending, or lexicographical order, and stored in different rows of matrix.
5. Next, similarities between pair of blocks are calculated using Euclidean distance [26] or Canberra distance [1] or Chebyshev distance [29], etc.
6. The pair of blocks for which similarity is less than a threshold value (a user-defined value) is removed, and the remaining pairs of blocks are highlighted with black or white.

Next, we present an overview of each category of region duplication detection method, as discussed above. Subsequently, we describe the operations of some of the representative methods from each of the above categories, in detail.

### *3.3.2 Dimensionality Reduction-Based Copy-Move Forgery Detection*

Here, we discuss the operating principle of three different methods, i.e., Principal Component Analysis (PCA) based method [16], the Singular Value Decomposition (SVD) based [8] method, and the combination of principal component analysis with discrete cosine transform-based method [21].

#### **3.3.2.1 Principal Component Analysis-Based Copy-Move Forgery Detection [16]**

Principal Component Analysis (PCA) is a statistical procedure to decorrelate a set of variables using some orthogonal transformation. It is generally used to remove the redundancies of statistical distribution and identify the principal components. In PCA-based region duplication detection method, the forged image of size $w \times h$ pixels is divided into overlapping blocks of size $B \times B$ pixels. The covariance ($C_m$) of each block is calculated as

$$C_m = \sum_{i=1}^{N_{total}} x_i x_i^T \tag{3.2}$$

where $x_i$ represents a block for $i = 1, 2, 3, \ldots, N_{total}$ and $N_{total} = (w - B + 1) \times (h - B + 1)$ represent the total number of fixed size overlapping blocks.

The principal components of $C_m$ are defined by the eigenvectors $e_j$ for $j = 1, 2, 3, \ldots, B$ (of $C_m$) corresponding to the eigenvalues $\lambda_j$ ($j = 1, 2, \ldots, b$ and $\lambda_1 > \lambda_2 > \cdots > \lambda_B$). Each image block can be linearly represented in terms of the eigenvectors, as

$$x_i = \sum_{j=1}^{B} a_j x_j \tag{3.3}$$

where $a_j = x_i^T e_j$ shows the new representation for each block. Each vector $x_i$ is truncated to first $N_t$ terms, $N_t$ being a user-defined parameter, in order to reduce the dimensionality of each block and generate a new $N_t$-dimensional representation of $C_m$, say $C_m^t$. The size of $C_m^t$ is $N_{total} \times B$.

To summarize, the steps of PCA-based copy-move forgery detection scheme utilizing $C_m^t$ are given below:

1. $C_m^t$ matrix is sorted row-wise and arrange in lexicographical order. A new matrix $S$ is obtained.
2. Let $s_i$ denote the $ith$ row of matrix $S$. The row $s_i$ of the matrix $S$ is represented using the tuple $(x_i, y_i)$, such that $(x_i, y_i)$ represents a block's image coordinates.
3. Every pair of rows $S_i$, $S_j$ where $|i - j| < N_n$ (user-defined value), is stored in a list L. Here, $N_n$ is the number of neighboring rows to be searched.
4. The offset frequency of each element in list $L$ is calculated as

$$\left.\begin{array}{lll} (x_j - x_i, y_j - y_i) & \text{if} & (x_i - x_j) < 0 \\ (x_i - x_j, y_i - y_j) & \text{if} & (x_i - x_j) > 0 \\ (x_i - x_j, y_i - y_j) & \text{if} & (x_i = x_j) \end{array}\right\} \tag{3.4}$$

5. The offset magnitude of each element in $S$ is calculated as

$$\sqrt{(x_i - x_j)^2 + (y_i - y_j)^2} \tag{3.5}$$

6. The block pairs with offset frequency smaller than $N_f$ (minimum frequency threshold) and offset magnitude smaller than $N_d$ (minimum offset threshold) are discarded. Here $N_f$ and $N_d$ are user-defined values.
7. The remaining pairs of blocks of the matrix $S$ are inferred to represent the copy-move blocks.

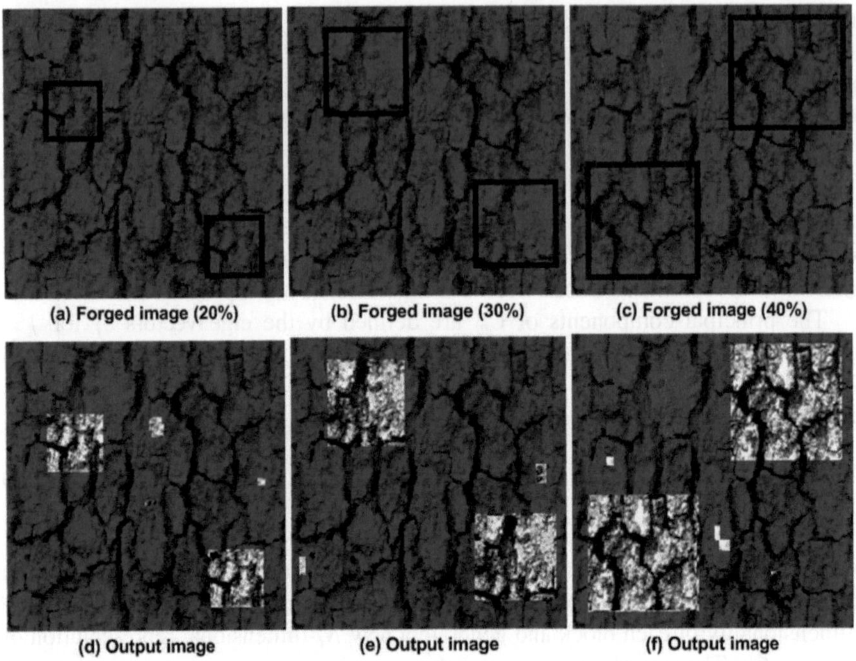


**Fig. 3.2** PCA-based region duplication detection: **a–c** forged images with varying forgery sizes; **d–f** copy-move forgery detected

An illustration of detection of copy-move forgery using the above method has been shown in Fig. 3.2. Figure 3.2a–c shows the forged image, and the forged area is varied to be 20, 30, and 40% of the entire image. The final output by the above algorithm has been shown in Fig. 3.2d–f.

#### 3.3.2.2 Singular Value Decomposition-Based Copy-Move Forgery Detection [8]

Singular Value Decomposition (SVD) is widely used in the field of pattern recognition, image and video compression, and signal processing. The image features extracted from SVD having algebraic invariant [10] property. Hence, features are more stable compared to PCA [10]. The steps involved in the detection of the copy-move duplication regions using SVD algorithm [8, 31] are given as follows:

1. A forged $w \times h$ image is divided into fixed size overlapping $B \times B$ blocks.
2. SVD is applied to each block and their corresponding singular values matrices $U$, $S$, and $V$ are extracted, expressed as

$$A = USV^T \tag{3.6}$$

where $U$ and $V$ are orthogonal matrices, and $S$ is a diagonal singular value matrix of the form:

$$S = \begin{bmatrix} \sigma_1 & 0 & & \cdots & & 0 \\ 0 & \sigma_2 & & \cdots & & 0 \\ 0 & 0 & \sigma_3 & \cdots & & 0 \\ & & & \vdots & & \\ 0 & 0 & 0 & \cdots & \sigma_r & 0 \\ 0 & 0 & 0 & \cdots & & 0 \end{bmatrix} \tag{3.7}$$

3. The diagonal elements of singular matrix, i.e., $\sigma_1, \sigma_2, \sigma_3, \ldots$ are the features of the block $B$ that are stored in a row in the feature matrix.
4. The feature vectors of the blocks that are stored row-wise in a matrix called the feature matrix, arranged into lexicographical order.
5. The Euclidean distances $D(u, v)$ between two rows, $u$ and $v$, of the feature matrix, are calculated as

$$D(u, v) = \sqrt{\sum_{i=1}^{r} (u_i - v_i)^2} \tag{3.8}$$

   where $u = (u_1, u_2, \ldots, u_r)$ and $v = (v_1, v_2, \ldots, v_r)$.
6. All the block pairs, for which $D(p, q) > T_1$ are eliminated. In the next step, further verification is executed on the remaining blocks.
7. The *Chebyshev distance* between two blocks $u$ and $v$ is calculated. Let $u$ and $v$ are the image blocks having coordinates $(i, j)$ and $(k, l)$, respectively. The Chebyshev distance is calculated as

$$C_{uv} = \max |abs(i - k), abs(j - l)| \tag{3.9}$$

8. If $C_{uv} \geq T_s$ then blocks $u$ and $v$ are marked as duplicate blocks, where $T_s$ is a user-defined value.

The detection of copy-move forgery using the above method has been shown in Fig. 3.3. Figure 3.3a–c shows the forged images, where forgery is varied to cover 20, 30, and 40% of the entire image. The final output by the above algorithm has been shown in Fig. 3.3d–f.

#### 3.3.2.3 PCA-DCT-Based Copy-Move Forgery Detection [21]

In [21] author proposed a region duplication detection algorithm combining the concept of PCA and DCT. In this method, the features of forged image are extracted by using DCT method, i.e., features are stored in the form of quantized coefficient in feature matrix. Dimensionality of the feature matrix is reduced by applying PCA. The PCA-DCT-based copy-move forgery detection scheme is comprised of the following steps:

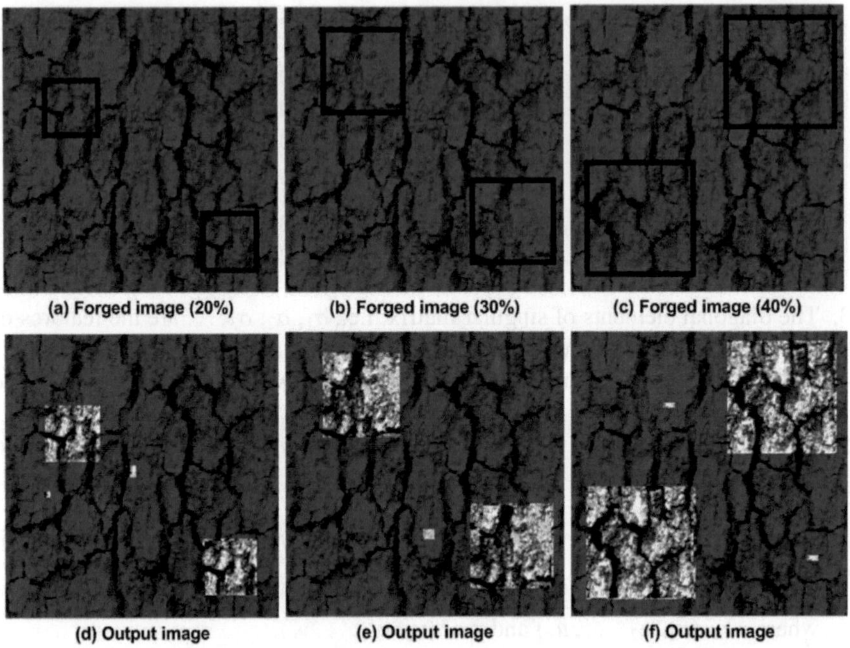


**Fig. 3.3** SVD-based region duplication detection: **a–c** forged images with varying forgery sizes; **d–f** images showing copy-move forgery detected

1. Here, we compute the quantized coefficients of each overlapping $B \times B$ block of a forged image, using DCT, which are placed row-wise into a feature matrix $M$.
2. We deal with only the first $\lceil q \times B^2 \rceil$ coefficients from every row of the feature matrix, where $q \in (0, 1)$. Hence we consider only a $\{(w - B + 1)(h - B + 1)\} \times \lceil q \times B^2 \rceil$ matrix, for further processing.
3. The size of feature matrix $M$ is reduced by PCA.
4. The rows of feature matrix $M$ are sorted lexicographically. The duplicate image blocks (similar rows) are located in $M$.
5. Let us suppose $(i_1, i_2)$ and $(j_1, j_2)$ are the coordinates of two similar blocks than the movement vector or shift vector $M_v$ for a pair of similar blocks is calculated as

$$M_v = (m_{v1}, m_{v2}) = (i_1 - j_1, i_2 - j_2) \tag{3.10}$$

6. For further processing, we consider the absolute values of movement vectors' $|M_v|$. The movement vectors $-M_v$ and $+M_v$ represent the same movement.
7. A frequency counter $C$ is used for calculation of similar blocks occurrence. The frequency counter $C$ is initialized to zero. For every pair of similar blocks, the frequency counter $C$ is increased by one: $C(m_{v1}, m_{v2}) = C(m_{v1}, m_{v2}) + 1$.
8. The frequency counter values corresponding to sift vector $M_{v1}, M_{v2}, \ldots$ of similar blocks are computed.

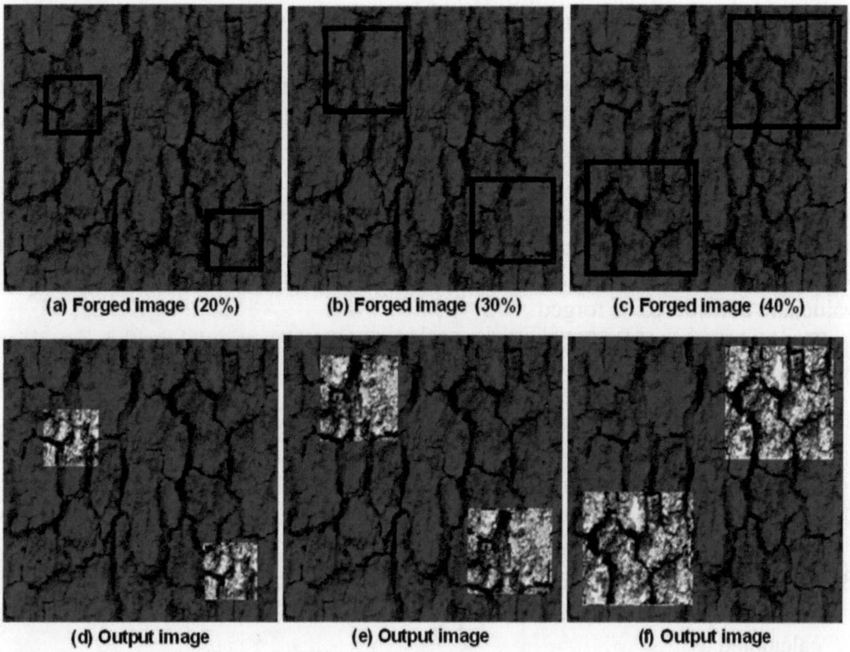


**Fig. 3.4** PCA-DCT-based region duplication detection: **a–c** forged images with varying forgery sizes; **d–f** copy-move forgery detected

9. Finally, at the end of the matching process, the identical blocks are found by the following condition:

$$C(M_v) > T \tag{3.11}$$

where $T$ is a manual threshold value.

The detection of copy-move forgery using the above method has been shown in Fig. 3.4. Figure 3.4a–c shows the forged image; the forgery is varying as 20, 30, and 40% of the entire image. The final output by above algorithm has been shown in Fig. 3.4d–f.

### 3.3.3 Discrete Cosine Transform-Based Copy-Move Forgery Detection

In this category, the features are stored in the form of quantized coefficients, extracted through discrete cosine transform, which contains the maximum information of image within the small number of coefficients. Here, we discuss the three existing Discrete Cosine Transform (DCT) based region duplication detection methods [6, 7, 25].

#### 3.3.3.1 Region Duplication Detection Using DCT [6]

In DCT-based copy-move forgery detection algorithm, initially if the test image is a color one, it is converted into gray scale. Next, the gray scale image is divided into fixed size overlapping blocks. From each block, we extract the quantized coefficient (features) by applying the DCT on each block. These quantized coefficients are sorted lexicographically and put into row of a matrix. Matching blocks are found using normalized *movement vectors* among all rows of matrix. The matching blocks, for which the value of the normalized movement vector is greater than the user-defined value, are considered as forged.

The basic steps of DCT-based copy-move forgery detection algorithm are given below:

1. Similar to the previous approaches, here also we divide a forged image into equal sized overlapping blocks and compute the quantized DCT coefficients of each block, which are stored into a feature matrix $M$.
2. Feature matrix $M$ is sorted lexicographically. Matching blocks or similar rows are located in the sorted feature matrix.
3. Let us suppose $(i_1, i_2)$ and $(j_1, j_2)$ are the coordinates of two similar blocks, and then the movement vector or shift vector $M_v$ for a pair of similar blocks is calculated as
$$M_v = (m_{v1}, m_{v2}) = (i_1 - j_1, i_2 - j_2) \tag{3.12}$$
4. We consider the absolute values of movement vector $|M_v|$, i.e., the movement vectors $-M_v$ and $+M_v$ represent the same movement.
5. A frequency counter, i.e., $C$ is used for calculation of similar block occurrence. The frequency counter $C$ is initialized with zero. For every pair of similar blocks, the frequency counter $C$ is increased by one: $C(m_{v1}, m_{v2}) = C(m_{v1}, m_{v2}) + 1$.
6. The frequency counter values corresponding to sift vector $M_{v1}, M_{v2}, \ldots$ of similar blocks are computed.
7. Finally at the end of the matching process, the identical blocks are found by the following condition:
$$C(M_v) > T \tag{3.13}$$
where $T$ is a manually determined threshold value.

The detection of copy-move forgery using the above method has been shown in Fig. 3.5. Figure 3.5a–c shows the forged image; the forgery is varying as 20, 30, and 40% of the entire image. The final output by above algorithm has been shown in Fig. 3.5d–f.

#### 3.3.3.2 Improved Copy-Move Forgery Detection Method Using DCT [7]

In [7], the authors have proposed an improved DCT based region duplication detection method. In this method, the forged image is divided into overlapping blocks

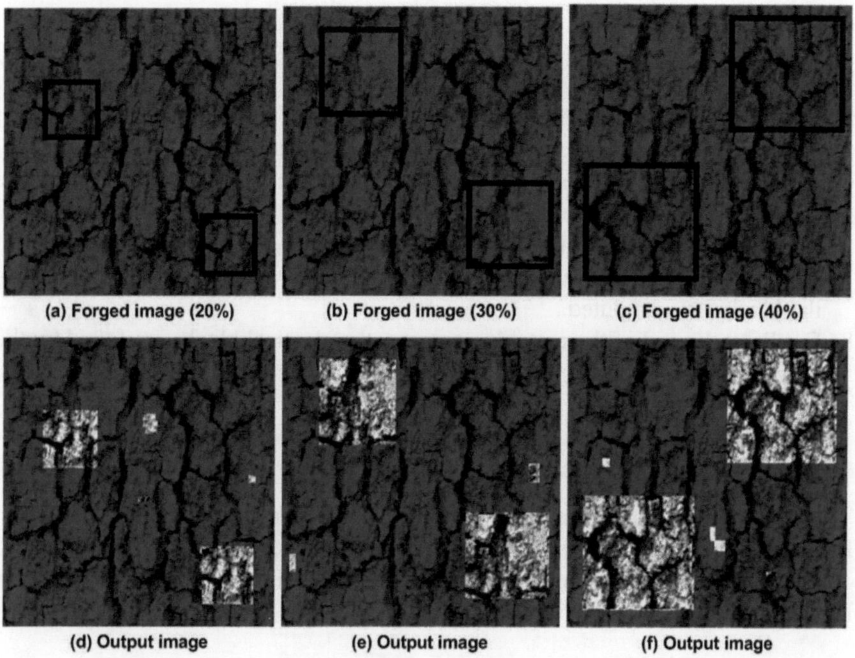


**Fig. 3.5** DCT-based region duplication detection: **a–c** forged images with varying forgery sizes; **d–f** images with copy-move forgery detected

and the quantized DCT coefficients are extracted. Initially, a factor $q$ $(0 < q < 1)$ is defined, and only the first $\lceil q \times B^2 \rceil$ quantized coefficients of each block are considered here. Due to that, the computational complexity of this method is low as compared to the previous method. The basic steps of improved DCT-based algorithms are given below:

1. In this approach, after storing the quantized image block DCT coefficients into feature matrix $M$, we select first $\lceil q \times B^2 \rceil$ coefficients from every row of the matrix, where $q \in (0, 1)$, for further processing. By this step, we generate a matrix $S$ having $(w - B + 1) \times (h - B + 1)$ rows and $\lceil q \times B^2 \rceil$ columns, which is sorted lexicographically.
2. Next, we compute the similarity among rows of the sorted feature matrix, by the following:
$$d = \sqrt{(x_i - x_j)^2 + (y_i - y_j)^2} \tag{3.14}$$
where$(x_i, y_i)$ and $(x_j, y_j)$ are the coordinate of two similar blocks.
3. The similar block pairs for which $d > N_d$, $N_d$ being another user-defined threshold are considered for further process.
4. Let us suppose $(i_1, i_2)$ and $(j_1, j_2)$ are the coordinates of two similar blocks, and then the movement vector or shift vector $M_v$ for a pair of similar blocks is

calculated as

$$M_v = (m_{v1}, m_{v2}) = (i_1 - j_1, i_2 - j_2) \tag{3.15}$$

5. We consider the absolute values of movement vectors $|M_v|$, i.e., the movement vectors $-M_v$ and $+M_v$ represent the same movement.
6. A frequency counter $C$ is used for calculation of similar block occurrence. The frequency counter $C$ is initialized with zero. For every pair of similar blocks, the frequency counter $C$ is increased by one: $C(m_{v1}, m_{v2}) = C(m_{v1}, m_{v2}) + 1$.
7. The frequency counter values corresponding to shift vector $M_{v1}, M_{v2}, \ldots$ of similar blocks are computed.
8. Finally at the end of the matching process, the identical blocks are found by the following condition:

$$C(M_v) > T \tag{3.16}$$

where $T$ is a manual threshold value.

The detection of copy-move forgery using the above method has been shown in Fig. 3.6. Figure 3.6a–c shows the forged image; the forgery is varying as 20, 30, and 40% of the entire image. The final output by the above algorithm has been shown in Fig. 3.6d–f.

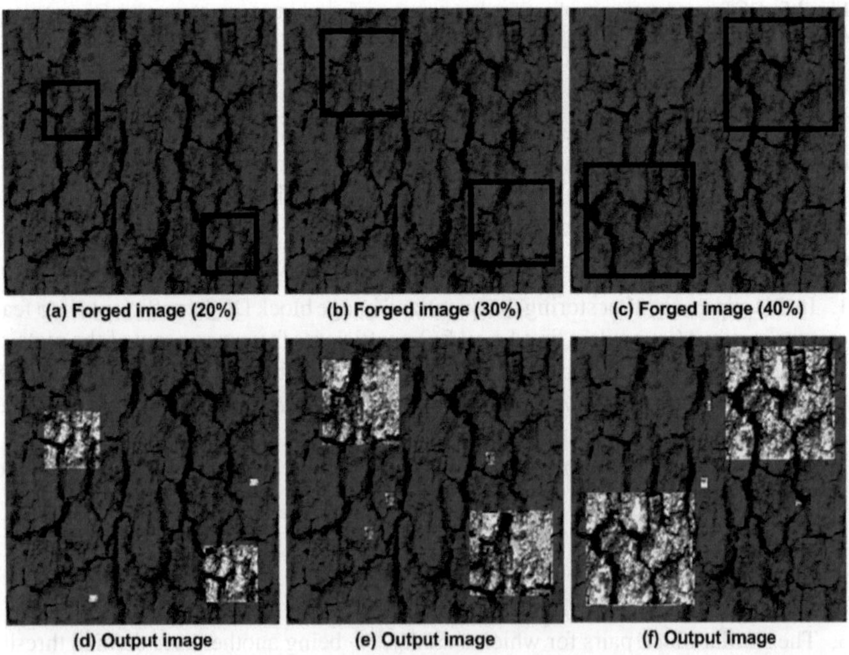


**Fig. 3.6** Improved DCT-based region duplication detection: **a–c** forged image with varying forgery sizes; **d–f** copy-move forgery detected

#### 3.3.3.3 An Efficient Copy-Move Forgery Detection Method Using DCT [25]

Wang et al. [25] developed another DCT-based copy-move forgery detection method. In this method, the forged image is divided into fixed size overlapping blocks and the DCT quantized coefficients are extracted. The mean of the DCT coefficients of each block is used as feature of image blocks. The basic steps of this method can be given as follows:

1. In this approach, for feature extraction, each $8 \times 8$ image block is divided into sixteen $2 \times 2$ nonoverlapping sub-blocks, and mean of the DCT coefficients is calculated for each sub-block. The array of 16 means serves as the block feature vector and is stored into a *package*.
2. For matching of duplicate regions, we calculate the pixel mean of each $(i, j)th$ image block as

$$P_{ij} = \frac{\sum_{x=1}^{8} \sum_{y=1}^{8} f_{ij}(x, y)}{64} \tag{3.17}$$

   where $f_{ij}(x, y)$ represents the intensity of the $(x, y)th$ pixel of the $(i, j)th$ image block.
3. *Packages* are produced to store DCT mean feature vectors along with the coordinates of the image blocks. Within the intensity range [0, 255] of 8-bit gray scale images, 64 packages are formed, each of offset 4. Distribution of block feature vectors into packages is done by the following principle. Features and coordinates of the $(i, j)th$ image block is put into the $\lfloor \frac{P_{ij}+1}{4} \rfloor th$ package.
4. Similar blocks within the same package are compared by their DCT mean features, to determine duplicates.
   To eliminate false matches, the mutual positions of the block pairs within a package are investigated, through shift vectors. Let us suppose $(i_1, i_2)$ and $(j_1, j_2)$ are the coordinate of two similar blocks $B_1$ and $B_2$, and then the movement vector or shift vector $M_v$ for a pair of similar blocks is calculated as

$$M_v = (m_{v1}, m_{v2}) = (i_1 - j_1, i_2 - j_2) \tag{3.18}$$

5. We consider the absolute values of movement vector $|M_v|$, i.e., the movement vectors $-M_v$ and $+M_v$ represent the same movement.
6. A frequency counter $C$ is used for calculation of similar block occurrence. The frequency counter $C$ is initialized with zero. For every pair of similar blocks, the frequency counter $C$ is increased by one: $C(m_{v1}, m_{v2}) = C(m_{v1}, m_{v2}) + 1$.
7. The frequency counter values corresponding to shift vector $M_{v1}, M_{v2}, \ldots$ of similar blocks are computed.
8. Finally at the end of the matching process, the identical blocks are found by the following condition:

$$C(M_v) > T \tag{3.19}$$

   where $T$ is a manual threshold value.

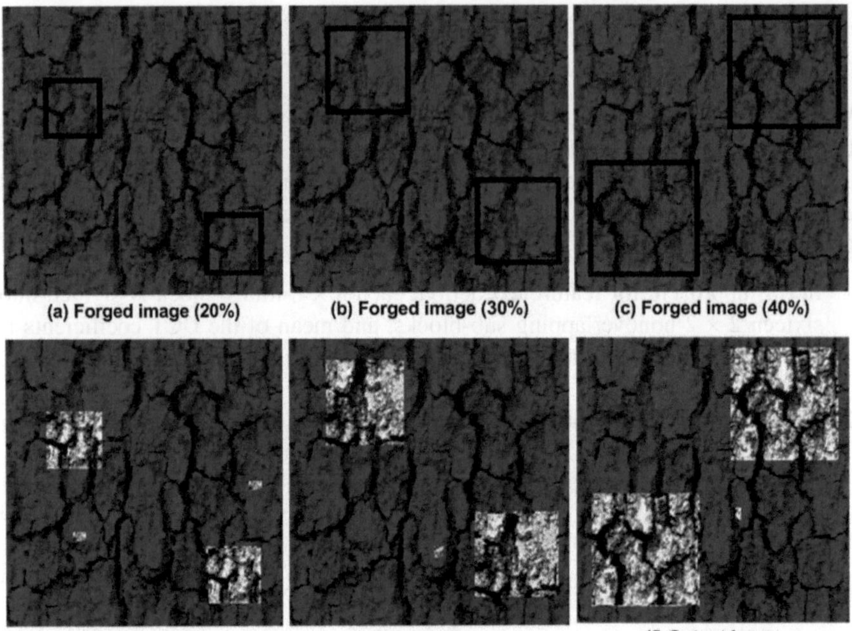


**Fig. 3.7** Efficient DCT-based region duplication detection: **a–c** forged image with varying forgery sizes; **d–f** copy-move forgery detected

The detection of copy-move forgery using the above method has been shown in Fig. 3.7. Figure 3.7a–c shows the forged image; the forgery is varying as 20, 30, and 40% of the entire image. The final output by above algorithm has been shown in Fig. 3.7d–f.

### 3.3.4 *Wavelet Transform-Based Copy-Move Forgery Detection*

Due to the multi-resolution property, wavelet transform has been very effective in several signal processing and vision based problems. Naturally, it has been found to be effective for digital image forensics applications also. In wavelet transform-based region duplication detection method, for feature extraction, we utilize wavelet transformations such as Discrete Wavelet Transform (DWT), Dyadic Wavelet Transform (DyWT), etc. Here, the image is decomposed into four sub-bands, out of four sub-band approximation sub-band contains the maximum amount of information, and hence plays the most important role in finding duplicate regions. The existing forgery detection methods related to this category introduce the dyadic wavelet transform-based method [15], dyadic wavelet transform with Zernike moment-based method [28], and the complex wavelet transform-based method [27].

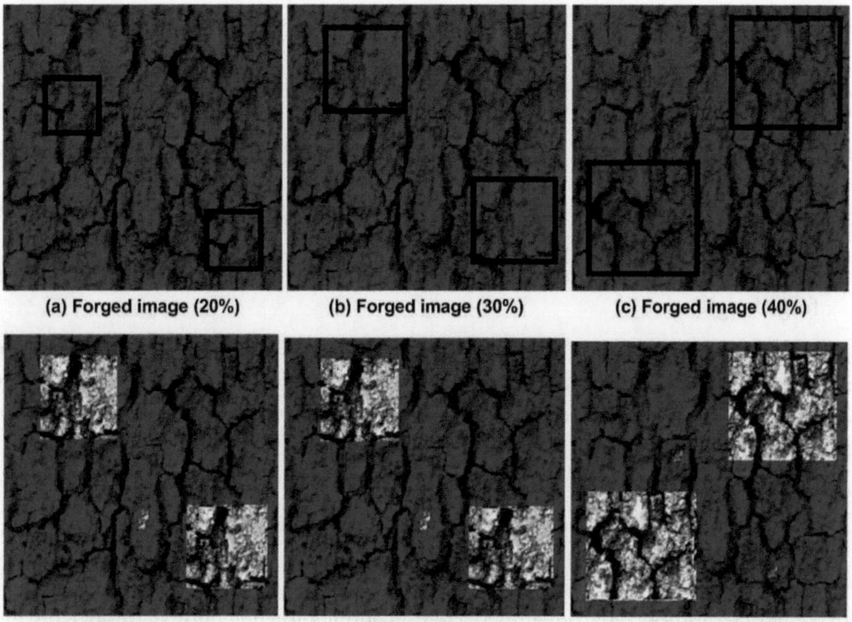


**Fig. 3.8** DyWT-based region duplication detection: **a–c** forged image with varying forgery sizes; **d–f** Copy-move forgery detected

#### 3.3.4.1 Copy-Move Forgery Detection Based on DyWT [15]

There exist many previous works on region duplication forgery detection using discrete wavelet transform. But due to the lack of shift invariance of DWT, such studies are not effective for shifting or combined types of attacks. To overcome the drawback, Mallat et al. [13] introduced the shift invariant version of DWT, i.e., Dyadic Wavelet Transform (DyWT). Appropriate low- and high-pass filters are applied to the signal at each level with no decimation to generate two sequences at the following level, each having the same length as the input sequence. The DyWT is a redundant scheme because the output obtained at each level of DyWT has an equal number of samples as the input. So, for an N-level decomposition, a redundancy of N is obtained in the wavelet coefficients. DyWT has been proven to give a better approximation than other wavelet transforms due to its inherent properties of redundancy.

In paper [15], authors have developed a DyWT-based copy-move forgery detection method. In detail, all the basic steps of this method are given below:

1. In this approach, DyWT is applied to the forged image, following which, the approximate and diagonal sub-bands are extracted.
2. Initially, the approximate sub-band is divided into overlapping blocks, the energy coefficients of which are saved. The similarity between a pair of blocks (say $p$ and $q$) is computed using Euclidean distance, as

$$D(p,q) = \sqrt{\frac{1}{N}\sum_{i=1}^{n}(p_i - q_i)^2}. \tag{3.20}$$

3. All the block pairs, for which $D(p,q) < T_1$ are considered, and arranged in increasing order in $list_1$, where $T_1$ is a user-defined threshold.
4. The process is repeated for the diagonal sub-band again. This time, the blocks satisfying the threshold criteria are arranged in $list_2$.
5. The blocks constituting at the same location in both the lists represent copy-move forged regions.

The detection of region duplication using the DyWT-based method has been shown in Fig. 3.8. Figure 3.8a–c shows the forged image; the forgery is varying as 20, 30, and 40% of the entire image. The final output by above algorithm has been shown in Fig. 3.8d–f.

#### 3.3.4.2 Duplication Detection Based on DyWT with Zernike Moment [28]

In paper [28], authors have developed a region duplication detection method using the operating principle of Dyadic Wavelet Transform (DyWT) and *Zernike moment* [9]. Here, initially, the forged image is decomposed by the DyWT and then features are extracted by Zernike moment.

The basic steps of DyWT with Zernike moment-based method are given below:

1. Similar to the previous approach, here also we obtain the DyWT sub-bands of a given image.
2. Initially, the approximation sub-band is divided into fixed size overlapping blocks of size $B \times B$ pixels, the total number of blocks produced being $(w - B + 1) \times (h - B + 1)$.
3. The features of each block are extracted using Zernike moment, as follows:

$$A(n,m) = \frac{n+1}{\prod} \int\int_{x^2+y^2\leq 1} f(x,y)(R_{nm}(\rho)exp\,(jm\theta))dxdy \tag{3.21}$$

where $R_{nm}(\rho)$ represents the Zernike polynomial, $n$ is the order of Zernike, $f(x,y)$ is image function, $\theta$ represents the angle between $\rho$ and $x$ axis, and $A(n,m)$ is the Zernike moment feature.

4. The magnitude of the Zernike moment of each block is stored into a feature matrix $M$, which is sorted lexicographically.
5. Similar blocks pairs or rows of the sorted matrix are computed using as

$$d = \sqrt{\sum_{k=0}^{n}(M_i(k) - M_{i+1}(k))^2} \tag{3.22}$$

where $1 \leq i < (w - B + 1) \times (h - B + 1)$.

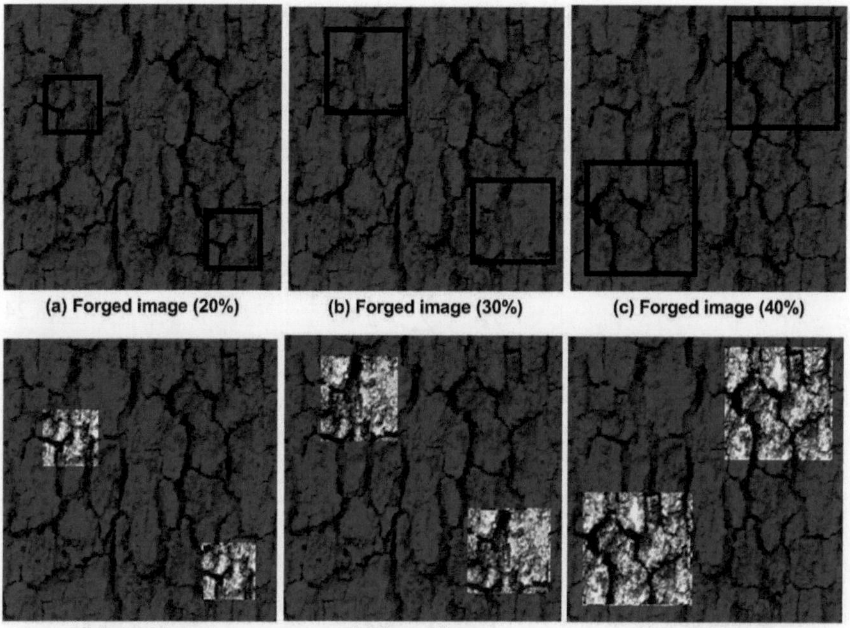


**Fig. 3.9** DyWT with Zernike moment-based region duplication detection. **a–c** Forged image with varying forgery sizes. **d–f** Copy-move forgery detected

6. Block pairs for which $d < T_1$ (where $T_1$ is a threshold of similarity of moments), are considered to be identical.
7. We optimize the false positives by calculation of the relative distances between pairs of blocks, as

$$d'(Block_i, Block_{i+1}) = \sqrt{(x_i - x_{i+1})^2 + (y_i - y_{i+1})^2} \tag{3.23}$$

where $(x_i, y_i)$ and $(x_{i+1}, y_{i+1})$ represent the image coordinates of $Block_i$ and $Block_{i+1}$, respectively.
8. All pairs of blocks for which $d'(Block_i, Block_{i+1}) > T_2$ ($T_2$ being user-defined value), are put into $list_1$.
9. The process is repeated for the diagonal sub-band, and the remaining block pairs are put into $list_2$.
10. Block pairs present at the same locations of the two lists ($list_1$ and $list_2$) represent the copy-moved blocks.

The detection of region duplication based on DyWT with Zernike moment has been shown in Fig. 3.9. Figure 3.9a–c shows the forged image; the forgery is varying as 20, 30, and 40% of the entire image. The final outputs by above algorithm have been shown in Fig. 3.9d–f.

#### 3.3.4.3 Complex Wavelet Transform-Based Copy-Move Forgery Detection [27]

Another wavelet transform-based copy-move forgery detection, proposed by Wu et al. [27], operates as follows:

1. Here, the forged image is divided into equal sized overlapping blocks, on which Complex Wavelet Transform (CWT) is applied, as

$$f_k^L = \sum_{d=0}^{6} E_d^L e^{\frac{-i\pi dk}{3}} \tag{3.24}$$

where $f_k^L$ represents feature extracted at level $L$, $k$ is the order of harmonic, and $E_d^L$ is the energy of the $dth$ sub-band at level $L$, computed as

$$E_d^L = \sum_{i=1}^{N} \sum_{j=1}^{N} ||x_d^L(i, j)|| \tag{3.25}$$

where each sub-band is $N \times N$ pixels, with complex coefficients $x_d^L(i, j)$ at level $L$. The extracted features are stored block-wise into rows of a feature matrix, which are arranged lexicographically.
2. Next, the similarity between two rows of the feature matrix is computed as

$$d = \sqrt{(x_i - x_j)^2 + (y_i - y_j)^2} \tag{3.26}$$

where $(x_i, y_i)$ and $(x_j, y_j)$ are the coordinates of two similar blocks in an image.
3. The similar block pairs for which $d > N_d$, a user-defined threshold parameter, are considered for further processing.
4. Let $(i_1, i_2)$ and $(j_1, j_2)$ be the coordinates of two similar blocks, and then the movement vector or shift vector $M_v$ for a pair of similar blocks is computed as

$$M_v = (m_{v1}, m_{v2}) = (i_1 - j_1, i_2 - j_2) \tag{3.27}$$

5. We consider the absolute values of movement vectors $|M_v|$. (The movement vectors $-M_v$ and $+M_v$ represent the same movement.)
6. A frequency counter $C$ is used for tracking of similar blocks occurrences, which is initialized to zero. For every occurrence of a similar block pair, the frequency counter $C$ is incremented by one: $C(m_{v1}, m_{v2}) = C(m_{v1}, m_{v2}) + 1$. Similarly, the frequency counter values corresponding to shift vector $M_{v1}, M_{v2}, \ldots$ of similar blocks are all computed.
7. At the end of the matching process, some blocks are finally output as identical blocks, based on an empirical threshold $T$, as

$$C(M_v) > T \tag{3.28}$$

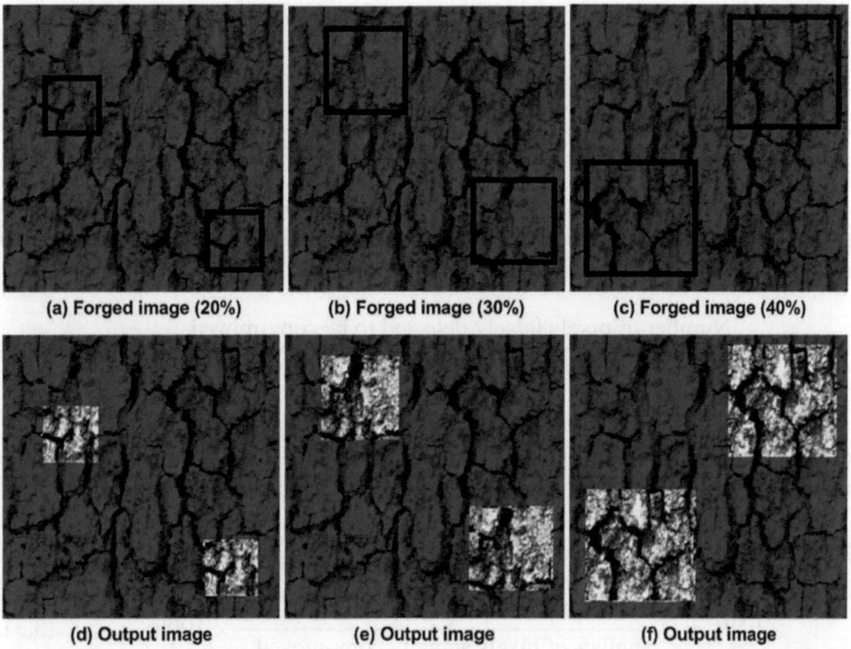


**Fig. 3.10** CWT-based region duplication detection: **a–c** forged image with varying forgery sizes; **d–f** copy-move forgery detected

In Fig. 3.10, the detection of duplicate regions using the above technique is shown. Figure 3.10a–c represents a test image, manually forged by us, where the forgery size is varied as 20, 30, and 40% of the entire image. The forgery detection results have been shown in Fig. 3.10d–f.

In the next section, we present a performance parameterization platform for evaluation, analysis, and comparison of existing block-based region duplication detection methods.

## 3.4 Three-Way Parameterization Platform

Here, we propose three parameters for evaluation of performance efficiency of copy-move image forgery detection algorithms, viz., *detection accuracy*, *false positive rate*, and *false negative rate*. On the basis of these proposed parameters, the accuracy of block-based copy-move forgery techniques may be estimated efficiently. The results of our evaluation are presented in Sect. 3.5. These parameters are defined on the basis of forged area in an image.

Detection Accuracy (DA) is proportional to correctly detected copy-moved pixels in forged image. Expressed as

$$DA = \frac{\text{Number of correctly detected copy-moved pixels}}{\text{Number of pixels actually copy-moved}} \times 100\% \quad (3.29)$$

False Positive Rate (FPR) is defined as the total number of non-manipulated image pixels, falsely detected to be manipulated, and expressed as a fraction of the total number of actually copy-moved pixels in the image. That is,

$$FPR = \frac{\text{Number of pixels falsely detected to be copy-moved}}{\text{Number of pixels actually copy-moved}} \times 100\% \quad (3.30)$$

The dyadic wavelet transform-based region duplication detection method, discussed in Sect. 3.3.4.1, applied to forged image, shown in Fig. 3.11a, produces false matches, which have been shown in Fig. 3.11c, with highlighted regions.

False Negative Rate (FNR) is the proportion to undetected copy-move pixels. Expressed as,

$$FNR = \frac{\text{Number of undetected copy-moved pixels}}{\text{Number of pixels actually copy-moved}} \times 100\% \quad (3.31)$$

The example of false negatives has been shown in Fig. 3.12. The dyadic wavelet transform-based region duplication detection method is applied to Fig. 3.12a, as an output the copy-move region is detected along with the false negatives shown in Fig. 3.12c.

On the basis of the proposed parameters, in the next section, we present the performance analysis of different block-based copy-move forgery detection techniques discussed in Sect. 3.3.

**Fig. 3.11** False positives or false matches in dyadic wavelet transform-based region duplication detection method. **a** Original image. **b** Forged image with copy-move forgery. **c** Output image (with false matches or false positives) marked with red

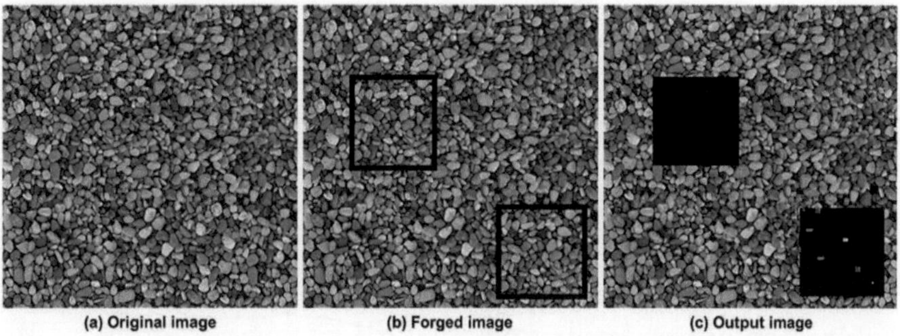


**Fig. 3.12** False negatives in region duplication detection method using dyadic wavelet transform algorithm. **a** Original image. **b** Forged image with copy-move forgery. **c** Detected copy-move forgery with false negatives

## 3.5 Experimental Results

### *3.5.1 Experimental Setup*

The block-based region duplication detection methods (discussed in Sect. 3.3) have been implemented in MATLAB 2014a, installed on a machine equipped with an Intel i7, 3.40 GHz Core 2 Duo processor and 4 GB of RAM. Here, we discuss the evaluation results for the block-based region duplication detection method discussed in Sect. 3.3. For our experiments, we used 70 standard test images (textured and natural) of size $256 \times 256$ pixels, collected from two image databases, i.e., CVG UGR Image Database [23] and USC-SIPI Image Database [20]. The test image is forged by software editing tool and forgery size is varied between 10 and 40% of test image.

In this section, on the basis of proposed parameters (i.e., detection accuracy, false positive rate, and false negative rate), we analyzed the performance characteristics of the discussed block-based algorithms related to all categories, i.e., dimensionality reduction-based copy-move forgery detection method, discrete cosine transform-based copy-move forgery detection methods and wavelet transform-based copy-move forgery detection method. We have also discussed the performance analysis with respect to the variation of block size and forgery size.

Finally, we interpret the experimental results so as to contribute entire guidance on merits and demerits of the existing methods discussed in this chapter.

### *3.5.2 Comparison of Detection Accuracy*

In terms of Detection Accuracy (DA), the performance comparison of different copy-move forgery detection methods, discussed in Sect. 3.3, has been shown in Fig. 3.13. In order to avoid experimental bias, all the results are taken as an average of overall forged images. According to Fig. 3.13, it is evident that as the forgery size increases,

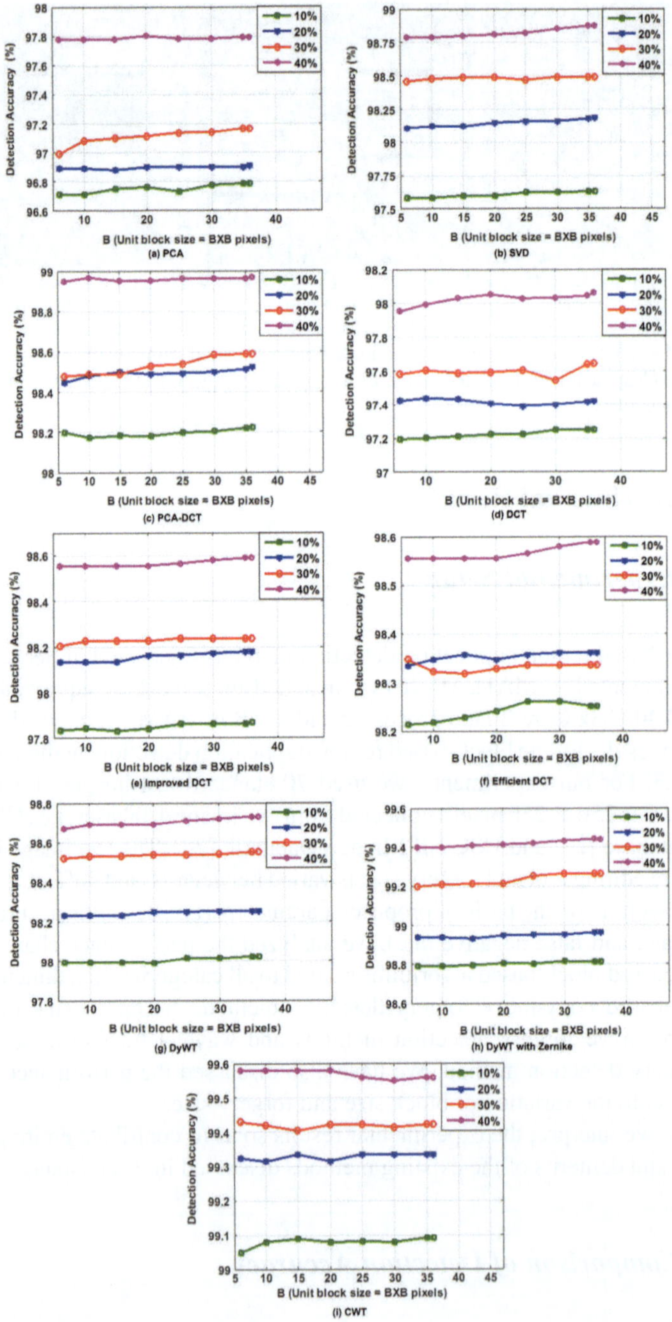


**Fig. 3.13** Performance characteristics of block-based copy-move forgery detection algorithms: Detection accuracy versus *B* and forgery size

**Table 3.1** Comparative results for detection accuracy

| Method | Forgery size (%) | DA (%) (maximum) | DA (%) (average) |
|---|---|---|---|
| PCA | 10 | 96.7870 | 96.6988 |
| | 20 | 96.9130 | 96.8309 |
| | 30 | 97.1671 | 97.0916 |
| | 40 | 97.7945 | 97.7645 |
| SVD | 10 | 97.6309 | 97.6092 |
| | 20 | 98.1880 | 98.1576 |
| | 30 | 98.4924 | 98.4754 |
| | 40 | 98.8730 | 98.8311 |
| PCA-DCT | 10 | 98.2219 | 98.1984 |
| | 20 | 98.5298 | 98.4812 |
| | 30 | 98.5945 | 98.5483 |
| | 40 | 98.9776 | 98.9520 |
| DCT | 10 | 97.2672 | 97.2254 |
| | 20 | 97.4396 | 97.4123 |
| | 30 | 97.6434 | 97.5978 |
| | 40 | 98.0624 | 98.0232 |
| Improved DCT | 10 | 97.8670 | 97.8121 |
| | 20 | 98.1810 | 97.1343 |
| | 30 | 98.2483 | 97.2099 |
| | 40 | 98.5882 | 98.5665 |
| Efficient DCT | 10 | 98.2692 | 98.2290 |
| | 20 | 98.3670 | 98.3480 |
| | 30 | 98.3494 | 98.3398 |
| | 40 | 98.5934 | 98.5607 |
| DyWT | 10 | 98.0229 | 98.0104 |
| | 20 | 98.2541 | 98.2387 |
| | 30 | 98.5592 | 98.5398 |
| | 40 | 98.7438 | 98.7096 |
| Zernike | 10 | 98.8179 | 98.8015 |
| | 20 | 98.9674 | 98.9329 |
| | 30 | 99.2662 | 99.2478 |
| | 40 | 99.4398 | 99.4199 |
| CWT | 10 | 99.0948 | 99.0739 |
| | 20 | 99.3493 | 99.3239 |
| | 30 | 99.4308 | 99.4193 |
| | 40 | 99.5893 | 99.5692 |

the detection accuracy increases for all the methods. The highest and average detection accuracy results of all the implemented methods are shown in Table 3.1.

Among all discussed block-based methods, the complex wavelet transform-based algorithm shows the highest detection accuracy, i.e., 99.59% for forgery size is 40% and the lowest detection accuracy is obtained for PCA-based copy-move forgery detection method, i.e., 96.6188% for forged size 10% and block size 36 × 36.

### 3.5.3 Comparison of False Positive Rate

Figure 3.14 reveals the comparison of the block-based copy-move forgery detection methods with respect to the false positive rate, with varying size of forgery and block size. From Fig. 3.14, it is observed that for all the methods, as the forgery size increases, the false positive rate decreases. False positive rate of wavelet transform-based region duplication detection methods is highest as compared to other categories. Moreover, many similar blocks get falsely detected at the boundaries of the test images; they can not be completely eliminated by adjusting the threshold. Table 3.2 shows the false positive rate, for all the algorithms, as varying forgery sizes. The false positive rate of DCT-based copy-move forgery detection is lowest, as evident from Table 3.2 and Fig. 3.14.

### 3.5.4 Comparison of False Negative Rate

Figure 3.15 and Table 3.3 show the false negative rate of discussed algorithms. In Fig. 3.15, the comparison of false negative rate versus block size and forgery size has been presented. From Fig. 3.15 it is evident that as the forgery size increases, the false negative rate diminishes for all the implemented algorithms. According to Table 3.3, it is observed that among all the block-based copy-move forgery detection techniques, complex wavelet transform-based algorithms exhibit the lowest false negative rate.

### 3.5.5 Trade-Off Between Detection Accuracy and Computational Complexity

The computational complexity of block-based region duplication detection method is dependent on the block size. Therefore, if the size of block is very small, then the total number of overlapping block will increase. Hence, the size of the feature matrix will be too large. Therefore, the computational complexity will increase. But according to our experimental result, smaller block size ensures higher detection accuracy. Hence, in such methods, it is desirable to obtain a reasonable trade-off between detection accuracy and computational complexity by selecting an appropriate block size.

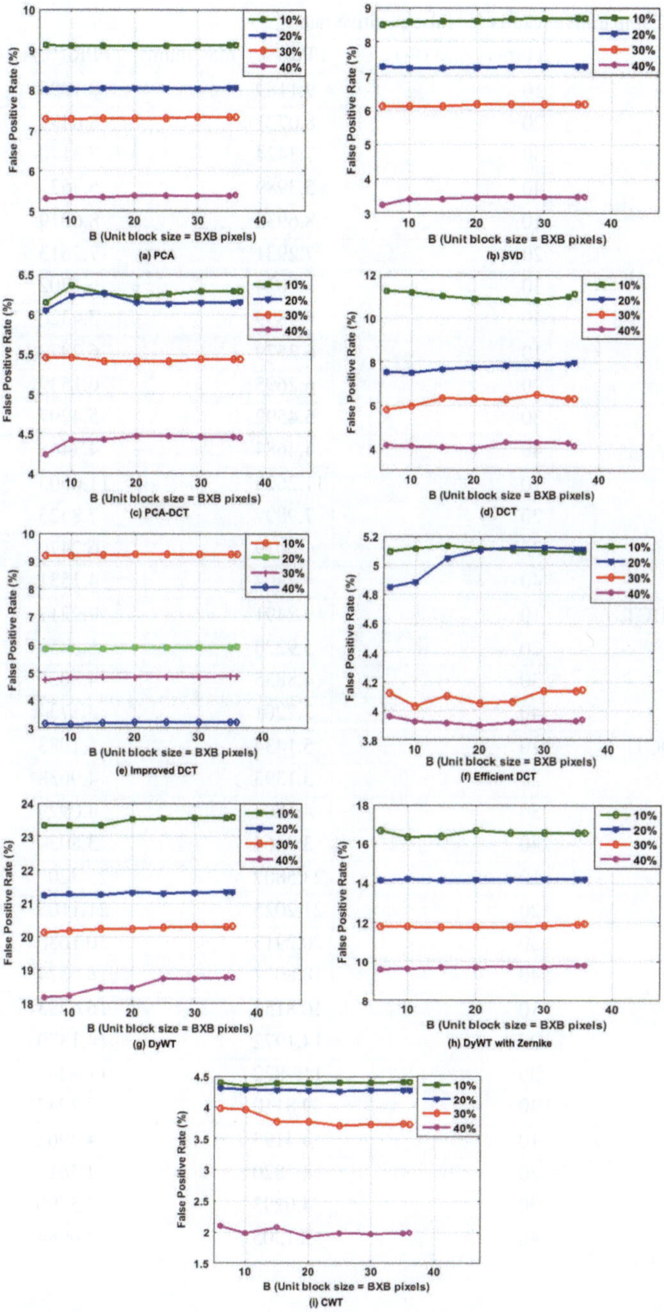


**Fig. 3.14** Performance characteristics of block-based copy-move forgery detection algorithms: False positive rate versus $B$ and forgery size

**Table 3.2** Comparative results for false positive rate

| Method | Forgery size (%) | FPR (%) (maximum) | FPR (%) (average) |
|---|---|---|---|
| PCA | 10 | 9.1189 | 9.1090 |
| | 20 | 8.0721 | 8.0495 |
| | 30 | 7.3424 | 7.3170 |
| | 40 | 5.3989 | 5.3637 |
| SVD | 10 | 8.6934 | 8.6019 |
| | 20 | 7.2931 | 7.2613 |
| | 30 | 6.2024 | 6.1702 |
| | 40 | 3.4899 | 3.4326 |
| PCA-DCT | 10 | 6.3579 | 6.2872 |
| | 20 | 6.2695 | 6.1539 |
| | 30 | 5.4592 | 5.4297 |
| | 40 | 4.4684 | 4.3468 |
| DCT | 10 | 11.2625 | 11.0503 |
| | 20 | 7.9897 | 7.8325 |
| | 30 | 6.5019 | 6.2470 |
| | 40 | 4.2864 | 4.1531 |
| Improved DCT | 10 | 9.2491 | 9.2217 |
| | 20 | 5.9223 | 5.8859 |
| | 30 | 4.8835 | 4.8036 |
| | 40 | 3.2201 | 3.1734 |
| Efficient DCT | 10 | 5.1434 | 5.1083 |
| | 20 | 5.1393 | 4.9638 |
| | 30 | 4.1804 | 4.0929 |
| | 40 | 3.9464 | 3.8934 |
| DyWT | 10 | 23.5802 | 23.3207 |
| | 20 | 21.2025 | 21.1105 |
| | 30 | 20.2913 | 20.1635 |
| | 40 | 18.8027 | 18.6374 |
| Zernike | 10 | 16.8155 | 16.6283 |
| | 20 | 14.1972 | 14.1370 |
| | 30 | 11.9822 | 11.9245 |
| | 40 | 9.8110 | 9.7047 |
| CWT | 10 | 4.4193 | 4.3962 |
| | 20 | 4.3820 | 4.3619 |
| | 30 | 4.0293 | 3.5790 |
| | 40 | 2.1305 | 2.0984 |

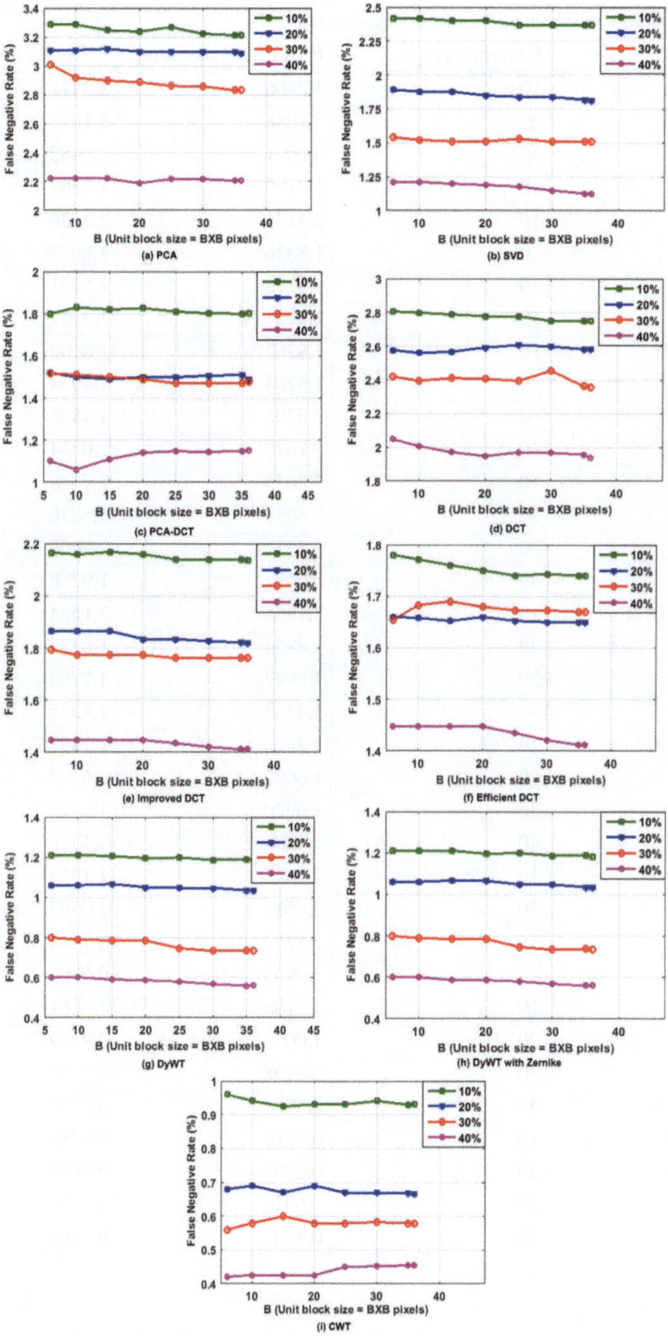


**Fig. 3.15** Performance characteristics of block-based copy-move forgery detection algorithms: false negative rate versus $B$ and forgery size

**Table 3.3** Comparative results for false negative rate

| Method | Forgery size (%) | FNR (%) (maximum) | FNR (%) (average) |
|---|---|---|---|
| PCA | 10 | 3.2866 | 3.2471 |
| | 20 | 3.1088 | 3.1016 |
| | 30 | 3.0601 | 2.8482 |
| | 40 | 2.2195 | 1.9367 |
| SVD | 10 | 2.4171 | 2.3236 |
| | 20 | 1.8626 | 1.8073 |
| | 30 | 1.5708 | 1.5171 |
| | 40 | 1.2221 | 1.1742 |
| PCA-DCT | 10 | 1.8267 | 1.8102 |
| | 20 | 1.5284 | 1.4938 |
| | 30 | 1.5203 | 1.4803 |
| | 40 | 1.1640 | 1.0834 |
| DCT | 10 | 2.8055 | 2.7758 |
| | 20 | 2.6083 | 2.5836 |
| | 30 | 2.4557 | 2.4007 |
| | 40 | 2.0477 | 1.9758 |
| Improved DCT | 10 | 2.1855 | 2.1624 |
| | 20 | 1.8854 | 1.6537 |
| | 30 | 1.8042 | 1.7703 |
| | 40 | 1.4447 | 1.4325 |
| Efficient DCT | 10 | 1.7820 | 1.7492 |
| | 20 | 1.6692 | 1.6484 |
| | 30 | 1.6920 | 1.6773 |
| | 40 | 1.4508 | 1.4378 |
| DyWT | 10 | 2.7443 | 1.1977 |
| | 20 | 2.8543 | 1.0507 |
| | 30 | 3.8234 | 0.7644 |
| | 40 | 3.3901 | 0.5808 |
| Zernike | 10 | 1.2108 | 1.3661 |
| | 20 | 1.0608 | 0.9279 |
| | 30 | 0.8008 | 0.7592 |
| | 40 | 0.6014 | 0.5826 |
| CWT | 10 | 0.9710 | 0.9387 |
| | 20 | 0.6992 | 0.6682 |
| | 30 | 0.6030 | 0.5802 |
| | 40 | 0.4602 | 0.4202 |

### 3.5.6 *Trade-Off Between Detection Accuracy and False Positive and Negative Rates*

From Tables 3.1, 3.2 and 3.3, it is evident that the wavelet transform-based algorithms have performed consistently well compared to other categories. Among all block-based region duplication detection methods, discussed in Sect. 3.3, the detection accuracy and false negative rate of complex wavelet transform-based algorithms are highest and lowest, respectively. But the false positive rate attained by this category is higher compared to the others. Moreover, the false positive rate of the Discrete Cosine Transform-based algorithm is lowest.

## 3.6 Conclusion

Region duplication or copy-move forgery is the most primitive forms of image forgery. Identifying the copy-move forgery is a challenging task, since there have been minimal changes in the statistical features of the original image due to this type of tampering. In the literature, several copy-move forgery detection schemes have been proposed. We have provided a detailed review and comparative study of such block-based copy-move forgery detection techniques. Moreover, a three-way parameterization technique has been proposed to choose the appropriate algorithm in the context of different types of forgeries. Therefore, as per user requirement, the presented parameterization may be helpful to select an appropriate method. Furthermore, the comparisons and claims are validated with rigorous experiments.

## References


1. Androutsos, D., Plataniotiss, K.N., Venetsanopoulos, A.N.: Distance measures for color image retrieval. In: 1998 Proceedings International Conference on Image Processing, ICIP 98, vol. 2, pp. 770–774. IEEE (1998)
2. Bayram, S., Sencar, H.T., Memon, N.: An efficient and robust method for detecting copy-move forgery. In: 2009 IEEE International Conference on Acoustics, Speech and Signal Processing, ICASSP, pp. 1053–1056. IEEE (2009)
3. Chen, W., Shi, Y.Q., Su, W.: Image splicing detection using 2-d phase congruency and statistical moments of characteristic function. In: Security, Steganography, and Watermarking of Multimedia Contents, pp. 65050R (2007)
4. Dixit, R., Naskar, R.: Review, analysis and parameterisation of techniques for copy-move forgery detection in digital images. IET Image Process. **11**(9), 746–759 (2017)
5. Farid, H.: Image forgery detection. Signal Process. Mag. **26**(2), 16–25 (2009)
6. Fridrich, A.J., David Soukal, B., Lukáš, J.: Detection of copy-move forgery in digital images. In: Proceedings of Digital Forensic Research Workshop. Citeseer (2003)
7. Huang, Y., Wei, L., Sun, W., Long, D.: Improved dct-based detection of copy-move forgery in images. Forensic Sci. Int. **206**(1), 178–184 (2011)

8. Kang, X.B., Wei, S.M.: Identifying tampered regions using singular value decomposition in digital image forensics. In: 2008 International Conference on Computer Science and Software Engineering, vol. 3, pp. 926–930. IEEE (2008)
9. Khotanzad, A., Hong, Y.H.: Invariant image recognition by zernike moments. IEEE Trans. Pattern Anal. Mach. Intel. **12**(5), 489–497 (1990)
10. Klema, V., Laub, A.: The singular value decomposition: its computation and some applications. IEEE Trans. Autom. Control **25**(2), 164–176 (1980)
11. Lee, J.-S.: Digital image enhancement and noise filtering by use of local statistics. In: IEEE Trans. Pattern Anal. Mach. Intel. PAMI **2**(2):165–168 (1980)
12. Lin, H.-J., Wang, C.-W., Kao, Y.-T., et al.: Fast copy-move forgery detection. WSEAS Trans. Signal Process. **5**(5), 188–197 (2009)
13. Mallat, S., Zhong, S.: Characterization of signals from multiscale edges. IEEE Trans. Pattern Anal. Mach. Intel. **14**(7), 710–732 (1992)
14. Masters, B.R., Gonzalez, R.C., Woods, R.: Digital image processing. J. Biomed. Optics **14**(2):029901 (2009)
15. Muhammad, G., Hussain, M., Bebis, G.: Passive copy move image forgery detection using undecimated dyadic wavelet transform. Digital Investig. **9**(1), 49–57 (2012)
16. Popescu, A.C., Farid, H.: Exposing digital forgeries by detecting duplicated image regions. Department Computer Science, Dartmouth College, Technology Report tr2004–515 (2004)
17. Redi, J.A., Taktak, W., Dugelay, J.L.: Digital image forensics: a booklet for beginners. Multimed. Tools Appl. **51**(1), 133–162 (2011)
18. Savchenko, V., Kojekine, N., Unno, H.: A practical image retouching method. In: 2002 Proceedings of First International Symposium on Cyber Worlds, pp. 480–487. IEEE (2002)
19. Sencar ... Overview of state-of-the-art in digital image forensics. Algorithms, Archit. Inf. Syst. Secur. **1**(3), 325–348 (2009)
20. SIPI. Image dataset. http://sipi.usc.edu/database/database.php?volume=misc
21. Sunil, K., Jagan, D., Shaktidev, M.: DCT-PCA based method for copy-move forgery detection. In: ICT and Critical Infrastructure: Proceedings of the 48th Annual Convention of Computer Society of India-Vol II, pp. 577–583. Springer (2014)
22. Tariang, D.B., Roy, A., Chakraborty, R.S., Naskar, R.: Automated jpeg forgery detection with correlation based localization. In: 2017 IEEE International Conference on Multimedia & Expo Workshops (ICMEW), pp. 226–231. IEEE (2017)
23. University of Granada. Color test images dataset. http://decsai.ugr.es/cvg/dbimagenes/c256.php
24. Wallace, G.K.: The JPEG still picture compression standard. IEEE Trans. Consum. Electron. **38**(1), xviii–xxxiv (1992)
25. Wang, H., Wang, H., Shi, C.: An efficient passive authentication scheme for copy-move forgery based on DCT. In: International Conference on Cloud Computing and Security, pp. 417–429. Springer (2016)
26. Wang, L., Zhang, Y., Feng, J.: On the euclidean distance of images. IEEE Trans. Pattern Anal. Mach. intel. **27**(8), 1334–1339 (2005)
27. YunJie, W., Deng, Y., Duan, H.B., Zhou, L.N.: Dual tree complex wavelet transform approach to copy-rotate-move forgery detection. Sci. China Inf. Sci. **57**(1), 1–12 (2014)
28. Yang, J., Ran, P., Xiao, D., Tan, J.: Digital image forgery forensics by using undecimated dyadic wavelet transform and zernike moments. J. Comput. Inf. Syst. **9**(16), 6399–6408 (2013)
29. Yap, P.T., Raveendran, P.: Image focus measure based on chebyshev moments. IEE Proc. Vis. Image Signal Process. **151**(2), 128–136 (2004)
30. Zhang, J., Feng, Z., Su, Y.: A new approach for detecting copy-move forgery in digital images. In: 2008 11th IEEE Singapore International Conference on Communication Systems, ICCS, pp. 362–366. IEEE (2008)
31. Zhang, T., Wang, R.-D.: Copy-move forgery detection based on svd in digital image. In: 2009 2nd International Congress onImage and Signal Processing, CISP'09, pp. 1–5. IEEE (2009)

# Chapter 4
# Copy-Move Forgery Detection Exploiting Statistical Image Features

## 4.1 Introduction

In this chapter, our aim to detect copy-move image forgery, where a region of an image is copied and moved to a different position within the same image, with the malicious motive of hiding or repeating important objects in the image. Many times attackers apply some post-processing operations (such as blurring, edge smoothing, noise addition and filtering of the forged region, etc.) on the forged image, so that the manipulated regions seem natural. Figure 4.1 shows a typical example of region duplication forgery, where a car object has been copied from the original image and pasted onto itself at a different location, to create the forged image. In recent years, many researchers have developed various digital forensic methods for copy-move forgery detection [1, 4–6, 8, 9, 13, 14]. Most of those techniques try to find efficient image features which would help to detect this form of forgery accurately.

Here, we present a region duplication detection method for images, which utilizes statistical image features in the frequency domain, viz., *mean* and *variance* [3]. The primary motivation is to exploit the local statistics of an image, using block-based mean and variance computations. The contribution of each individual block with respect to the entire image, in terms of intensity, can be measured using the image mean, whereas the variation of individual pixel intensities within an image sub-block, is indicated by the variance. First, the forged image is decomposed into four sub-bands using discrete wavelet transform. Next, the approximate sub-band is split into fixed size overlapping blocks. The statistical image features are extracted using the mean value of each block and stored into a matrix. The similarities between the rows of matrix are calculated using Euclidean distance. Copy-move blocks are decided by using user-defined value. The forgery detection accuracy of this scheme is further improved, by minimizing the rate of false positives, through consideration of variance of duplicate image blocks. Detailed steps of the scheme are presented in Sect. 4.3.

The rest of the chapter is structured as follows. A review of related work has been provided in Sect. 4.2. In Sect. 4.3, we describe the proposed mean- and variance-based


A. Roy et al., *Digital Image Forensics*, Studies in Computational Intelligence 755, https://doi.org/10.1007/978-981-10-7644-2_4

**Fig. 4.1** Example of copy-move forgery

image forgery detection method in detail. In Sect. 4.4, we present the experimental results. The chapter is summarized in Sect. 4.5.

## 4.2 Related Work

Recently, many researchers have presented region duplication detection techniques for digital images, for example [1, 4–6, 8, 9, 13, 14]. In paper [4], the authors proposed exhaustive search, autocorrelation, DCT or robust matching, and exact block matching-based region duplication detection methods. Out of these, DCT-based method efficiently detects duplicate region comparison to others. In paper [9], the authors discussed a Principal Component Analysis (PCA) based methodology. This scheme is not performed well in case of lossy compressed image. In [2], authors implemented a discrete cosine transform-based algorithm for region duplication identification. This method is significantly effective to identify multiple copy-move forgeries in an image. In [10], the authors presented a copy-move forgery detection method based on curvelet transform. This technique is robust to geometric attacks with considerably low computational complexity as compared to other related works. In paper [5], authors presented a technique which is robust against lossy JPEG compression and additive Gaussian noise. In [1], the authors proposed an approach based on Fourier Mellin Transform (FMT), which is also capable of detecting copy-rotate-move (small angle) and copy-scale-move forgeries in addition to plain duplicate region identification. In [7], the authors propose a double quantization DCT-based scheme, which is robust against JPEG compression. Another noteworthy copy-move forgery detection algorithm, robust to JPEG compression was implemented by Muhammad et al. [8], which is based on undecimated Dyadic Wavelet Transform (DyWT).

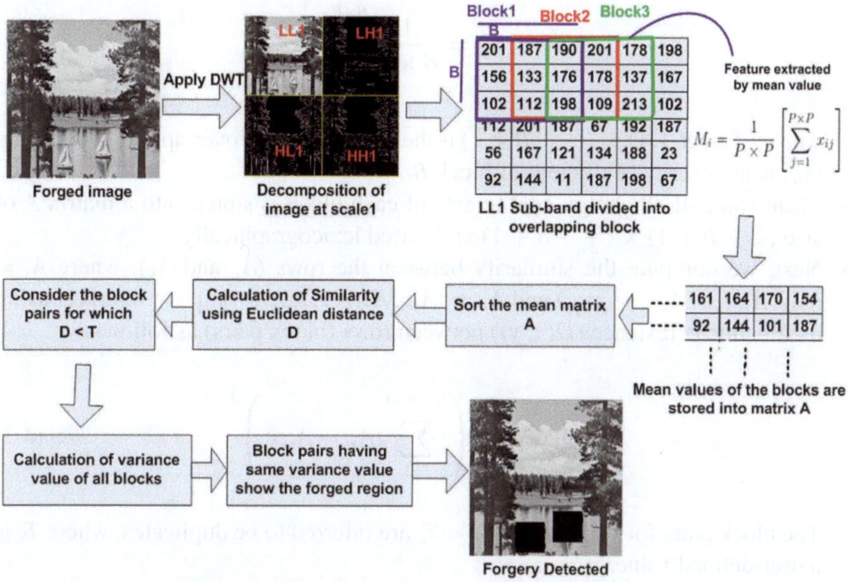


**Fig. 4.2** Operational steps of the proposed algorithm

## 4.3 Region Duplication Detection Technique Using Statistical Image Features

In this section, we present a region duplication detection algorithm, exploiting statistical features of an image, i.e., mean and variance, in DWT domain [3]. Figure 4.2 shows the flowchart of the presented technique. The steps of this technique constitute the followings:

1. Let us consider an input image of size $W \times H$ pixels. If the input image is a color one, we convert it into gray scale as

$$\text{Gray intensity} = 0.299 \times \text{Red intensity} + 0.587 \times \text{Green intensity} + 0.114 \times \text{Blue intensity} \tag{4.1}$$

2. DWT is applied to the image $I$ to obtain four sub-bands at scale 1, viz., approximation ($LL_1$), horizontal ($LH_1$), vertical ($HL_1$) and diagonal ($HH_1$). $LL_1$ contains the maximum information of the test image because it corresponds to low entropy as well as high similarity factor content. Hence, we have chosen $LL_1$ for next steps.
3. The $LL_1$ sub-band of size $\frac{W}{2} \times \frac{B}{2}$ is divided into overlapping blocks of size $B \times B$ pixels.
4. The mean value sequence $M_1, M_2, \ldots, M_{(\frac{W}{2}-P+1)\times(\frac{H}{2}-P+1)}$ is computed from the blocks $B_1, B_2, \ldots, B_{(\frac{W}{2}-B+1)\times(\frac{H}{2}-B+1)}$, as

$$M_i = \frac{1}{B \times B} \sum_{j=0}^{B \times B} x_{ij} \tag{4.2}$$

where $(\frac{W}{2} - B + 1) \times (\frac{H}{2} - B + 1)$ is the total number of overlapping blocks, and $(x_{ij})$ is the energy coefficient of block $B_i$.

5. Mean value of all energy coefficients of each block is stored into a matrix $A$ of size $(\frac{W}{2} - B + 1) \times (\frac{H}{2} - B + 1)$ and sorted lexicographically.
6. Next, we compute the similarity between the rows ($A_x$ and $A_y$), where $A_x = (A_{x1}, A_{x2}, \ldots A_{x \times (\frac{H}{2} - P + 1)})$ and $A_y = (A_{y1}, A_{y2}, \ldots A_{y \times (\frac{H}{2} - P + 1)})$ of matrix $A$ using the Euclidean distance $(D(x, y))$ between rows (block pairs) as follows:

$$D(x, y) = \left( \sum_{i=1}^{\frac{H}{2} - P + 1} (A_{xi} - A_{yi})^2 \right)^{\frac{1}{2}} \tag{4.3}$$

7. The block pairs for which $D(x, y) < T_s$ are inferred to be duplicates, where $T_s$ is a user-defined value.

Next, we present the proposed false match reduction scheme.

### 4.3.1 *Reducing False Matches*

Reduction of the false matches in copy-move forgery detection is a major requirement, especially given the widespread application of such techniques in the legal domain. It was observed that the proposed algorithm (discussed above) is capable of detecting duplicate regions within an image, but also results in false positives. Hence, to control these false positives we now present a variance based technique. Basics steps for reduction of false matches or false positives are given below:

1. The variance of each block pairs (obtained by step 6) is calculated as

$$V_i = \frac{1}{B \times B} \sum_{j=1}^{B \times B} (x_{ij} - M_i)^2 \tag{4.4}$$

2. All the block pairs having similar variances are considered as duplicates, i.e., blocks $B_i$ and $B_k$ are decided to be duplicates if and only if $V_i = V_k$. Rest all are decided to be authentic.

In Fig. 4.3, we provide the results of the above-discussed method when applied on two forged images. We also present in Fig. 4.3b, e the false matches resulting from this method. The final optimized output is presented in Fig. 4.3c, f. In the next section, we have presented detailed experimental results and associated discussions.

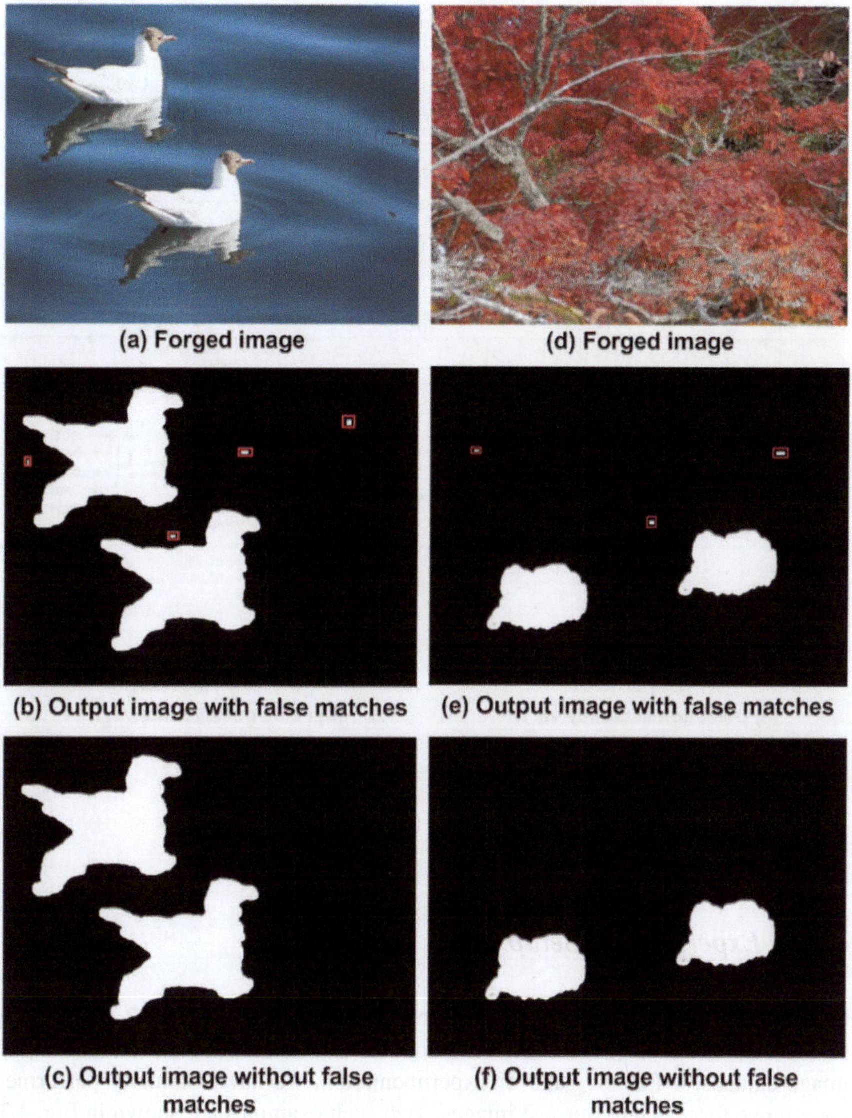


**Fig. 4.3** Output of the mean-variance-based method. Detected duplicate regions are shown in white, while the false matches are shown with red boundaries

**Table 4.1** Detection accuracy and false positive rate results for the presented region duplication detection technique

| Block size (pixels) | DA (%) | $FPR_1$ (%) | $FPR_2$ (%) |
|---|---|---|---|
| 5 × 5 | 98.5186 | 6.2592 | 2.7709 |
| 10 × 10 | 98.4263 | 6.1293 | 2.3956 |
| 15 × 15 | 98.4092 | 6.1129 | 2.4251 |
| 20 × 20 | 98.4007 | 6.0382 | 2.8134 |
| 25 × 25 | 98.3770 | 6.0647 | 2.4074 |
| 30 × 30 | 98.3089 | 6.0184 | 2.4692 |
| 35 × 35 | 98.2972 | 5.9374 | 2.6008 |

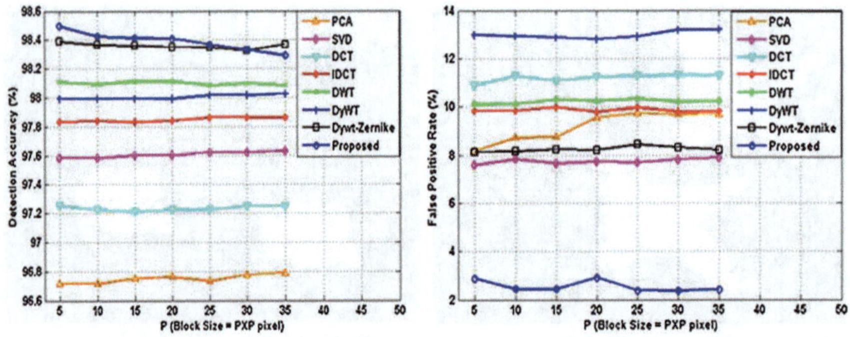


(a) Detection accuracy vs. P (b) False positive rate vs. P

**Fig. 4.4** Performance comparison results

## 4.4 Experimental Results

### *4.4.1 Experimental Setup*

Our test dataset consists of a set of different sized (128 × 128 and 512 × 512) gray scale as well as color images, collected from the CoMoFoD [11] and USC SIPI [12] image databases. For the sake of experimentation, we have manually performed copy-move forgery into our test images. Two such examples are shown in Fig. 4.3, where the test images have been manually forged.

### *4.4.2 Comparison and Discussion*

The Detection Accuracy (DA) and False Positive Rate (FPR) results (as defined in the previous chapter in Sect. 3.4) of the technique presented in Sect. 4.3 are presented in Table 4.1 (averaged over all test images). Column $FPR_2$ indicates the improvement in

FPR, achieved after utilizing the variance-based false matches reduction technique, discussed in Sect. 4.3.1. The performance analysis in terms of detection accuracy and false positive rate with variation in the block size is presented in Table 4.1. From Table 4.1, it can be inferred that *DA* is not strongly affected with variation in the block size. The detection accuracy lies between 98.29 and 98.51% as block size changes from $5 \times 5$ to $35 \times 35$. Initially, the *FPR* value lies between 5.94 and 6.26%, and after applying the variance based false positives reduction method, the false positive rate is reduced, i.e., 2.39–2.81% (depending on the block size).

We perform a comparison between the proposed technique and six different existing region duplication detection methods, viz., (a) Discrete Cosine Transform (DCT) [4]; (b) Singular Value Decomposition (SVD) [6]; (c) Dyadic Wavelet Transform (DyWT) [8]; (d) Principal Component Analysis (PCA) [9]; (e) dyadic wavelet transform with Zernike moment [13]; (f) Improved DCT [5]; and (g) Discrete Wavelet Transform (DWT) [14]. Figure 4.4 presents the comparison results in terms of detection accuracy versus *B*, and false positive rate versus *B* plots. From this figure, it is observed that the mean-variance-based technique attains higher detection accuracy and lower false positive rate than the existing methods considered here.

## 4.5 Conclusion

In this chapter, we present a region duplication forgery detection method utilizing mean and variance statistical features. Those are simple yet powerful statistical image features and effective for forgery detection also. The presented method operates by dividing an image into pixel sub-blocks in the frequency domain, and also deals with the statistical features of all image blocks. The experimental results verify the efficacy of the method and show the supremacy of the technique over the existing copy-move forgery detection techniques.

## References

1. Bayram, S, Sencar, H.T., Memon, N.: An efficient and robust method for detecting copy-move forgery. In: IEEE International Conference on Acoustics, Speech and Signal Processing, 2009. ICASSP 2009, pp. 1053–1056. IEEE (2009)
2. Cao, Y., Gao, T., Fan, L., Yang, Q.: A robust detection algorithm for copy-move forgery in digital images. Forensic Sci Int **214**(1), 33–43 (2012)
3. Dixit, R., Naskar, R., Sahoo, A.: Copy-move forgery detection exploiting statistical image features. In: 2017 International Conference on Wireless Communications, Signal Processing and Networking (WiSPNET), pp 2277–2281. IEEE (2017)
4. Jessica Fridrich, A., David Soukal, B., Jan Lukáš, A.: Detection of copy-move forgery in digital images. In: Proceedings of Digital Forensic Research Workshop. Citeseer (2003)
5. Huang, Y., Wei, L., Sun, W., Long, D.: Improved DCT-based detection of copy-move forgery in images. Forensic Sci Int **206**(1), 178–184 (2011)

6. Kang, X.B., Wei, S.M.: Identifying tampered regions using singular value decomposition in digital image forensics. In: 2008 International Conference on Computer Science and Software Engineering, vol. 3, pp. 926–930. IEEE (2008)
7. Lin, Z., He, J., Tang, X., Tang, C.-K.: Fast, automatic and fine-grained tampered JPEG image detection via DCT coefficient analysis. Pattern Recognit. **42**(11), 2492–2501 (2009)
8. Muhammad, G., Hussain, M., Bebis, G.: Passive copy move image forgery detection using undecimated dyadic wavelet transform. Dig. Investig. **9**(1), 49–57 (2012)
9. Popescu, A.C., Farid, H.: Exposing digital forgeries by detecting duplicated image regions. Department computer science, Dartmouth college, Technology report tr2004-515 (2004)
10. Qiao, M., Sung, A., Liu, A., Ribeiro, B.: A novel approach for detection of copy-move forgery. In: Proceedings of Fifth International Conference on Advanced Engineering Computing and Applications in Sciences (ADVCOMP 2011), pp. 44–47 (2011)
11. Tralic, D., Zupancic, I., Grgic, S., Grgic, M.: Comofod–new database for copy-move forgery detection. In: 2013 International symposium on ELMAR, pp. 49–54. IEEE (2013)
12. Weber, A.G.: The USC-SIPI image database version 5. USC-SIPI Report, vol. 315, pp. 1–24 (1997)
13. Yang, J., Ran, P., Xiao, D., Tan, J.: Digital image forgery forensics by using undecimated dyadic wavelet transform and zernike moments. J. Comput. Inf. Syst. **9**(16), 6399–6408 (2013)
14. Zhang, J., Feng, Z., Su, Y.: A new approach for detecting copy-move forgery in digital images. In: 11th IEEE Singapore International Conference on Communication Systems, 2008. ICCS 2008, pp. 362–366. IEEE (2008)

# Chapter 5
# Copy-Move Forgery Detection with Similar But Genuine Objects

## 5.1 Introduction

Copy-move image forgery is a popular and widely used tampering method in digital images. In this type of forgery, some parts of the pristine image are copied and pasted back onto the same image, with or without some form of manipulation (i.e., scaling, rotation, illumination change, etc.) to produce the forged image. The existing *Copy-Move Forgery Detection* (CMFD) techniques in the literature can be classified into two major categories: (a) block-based and (b) keypoint-based matching techniques. Fridrich et al. [8] have come up with the first block-based CMFD algorithm using *Discrete Cosine Transform* (DCT) and lexicographic ordering. Muhammad et al. [18] utilized *Dyadic Wavelet Transform* (DyWT) for CMFD using approximation and detail sub-band statistics. Dimensionality reduction-based approaches using *Singular Value Decomposition* (SVD) [13], *Principal Component Analysis* (PCA) [21], etc., have also been proposed in the literature. Besides, numerous techniques robust to geometric transformations have also been proposed using *Fourier Mellin Transform* (FMT) [14], *Zernike moment* [23], *multi-resolution local binary pattern* [7], etc.

However, the keypoint matching-based CMFD techniques are more efficient in terms of both performance and computational overhead. Huang et al. [12] proposed CMFD algorithm using *Scale-Invariant Feature Transform* (SIFT), using the presence of matched features as an indication of image forgery. Pan et al. [20] improvised the technique using affine geometric transformation estimation among the matched SIFT keypoints. Amerini et al. [2] further improvised the scheme using hierarchical clustering of the matched features, followed by homography based geometrical transformation estimation. Detection of forged image is done based on the number of geometrical transformations found. Christlein et al. [5] provides an excellent survey of popular CMFD approaches and also introduces a new copy-move forgery detection database, viz., "Manipulate". Computationally fast *Speeded-Up Robust Features* (SURF) have been used for forgery detection by Xu et al. [26]. Recently, several CMFD techniques have also been proposed using *DAISY* descriptor [10], *Harris*



A. Roy et al., *Digital Image Forensics*, Studies in Computational Intelligence 755, https://doi.org/10.1007/978-981-10-7644-2_5

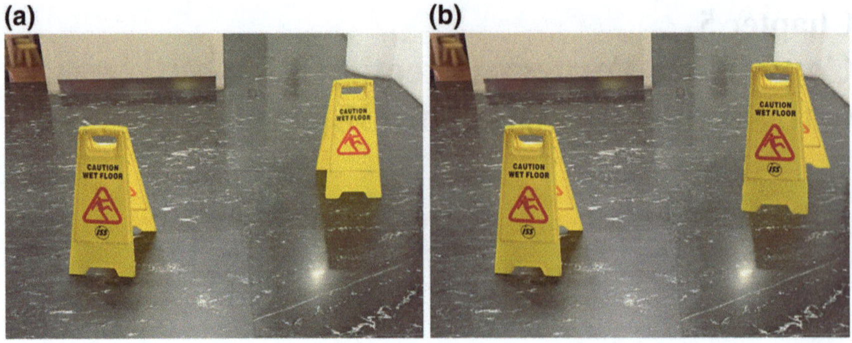


**Fig. 5.1** Example of copy-move forgery with SGO: **a** original image containing SGO and **b** forged image

*points* and *step sector statistics* [4], patchmark-based *Dense-Field Descriptor* [6], etc.

Existing keypoint and block-based CMFD algorithms consider region similarity as an indication of copy-move forgery, without considering the fact that the in practice scene might subsume *Similar but Genuine Objects* (SGO) as shown in Fig. 5.1. Since the origin of the forgery resides in the tampered image itself, in general, the CMFD problem can be simply solved by searching for similar region within the image. However, this in fact makes the forgery detection rather inconvenient for images having SGO. Wen et al. [25] introduced a novel dataset, viz., *COpy-moVe forgERy dAtabase with similar but Genuine objEcts* (COVERAGE) and observed that the performance of the state-of-the-art techniques degrades significantly for images containing SGO in the COVERAGE database. The reason being the fact that the state-of-the-art techniques find region similarity in both the pristine image with SGO, and also the tampered image. Hence, these techniques are prone to substantial performance degradation due to false positives, i.e., they infer the pristine image with SGO as tampered. This motivates us to pay attention to this interesting yet less explored variant of the CMFD problem. Approaching this problem with *Scaled Harris Feature Descriptor* (SHFD) Zhu et al. [28] obtained partial success. However, their approach lacks robustness against detecting complex tampering (i.e., illumination change, free-form, and combination of these). Moreover, their technique performs slightly better than the state-of-the-art SIFT [2], SURF [26], and dense-field [6] based approaches.

In this chapter, we describe the approach of using *Rotated Local Binary Pattern* (RLBP) features along with SURF keypoint detection for more accurate CMFD detection, with images having SGO [22]. The major insight behind applying the RLBP features is the fact that while performing copy-move image forgery, tampered object embedding introduces unnatural local artifacts at the boundaries of the tampered object. Such local artifacts rendered the tampered image more textured than the original image. Therefore, texture features, viz., RLBP can be effectively discriminative identifying tampered image and original image with SGO. Moreover, RLBP is a rotation-invariant texture feature. This justifies the efficacy of RLBP features for forgery detection over other state-of-the-art features. Experimental validation has been provided in the results section.

The rest of the chapter is organized as follows. The forgery detection framework has been discussed in Sect. 5.2, followed by the experimental validation in Sect. 5.3. Finally, conclusions are drawn in Sect. 5.4.

## 5.2 Proposed Method

The proposed forgery detection methodology [22] consists of four major steps as follows: (a) SURF keypoint detection, (b) RLBP feature extraction, (c) Feature matching, and finally, (d) hierarchical agglomerative clustering and forgery detection. We shall discuss each of the steps subsequently.

### *5.2.1 Keypoint Detection*

*Speeded-Up Robust Features (SURF)* is a well-known keypoint detector as well as descriptor proposed by Bay et al. [3]. It is very effective in many computer vision problems, viz., object recognition, classification, 3D reconstruction, image registration, etc. These features are partially motivated by the SIFT features proposed by Lowe [15]. However, it has certain advantages like it is faster and robust to certain image transformation compared to SIFT [3].

The keypoints are identified by using a "Fast-Hessian Detector", i.e., based on an approximation of the Hessian matrix of a given image point as described in [3]. An example of the SURF keypoints and descriptors are shown in Fig. 5.2.

In SIFT, Laplacian of Gaussian (LoG) filter has been approximated with Difference of Gaussian (DoG) filter for finding scale-space. However, for SURF, LoG has been approximated with Box Filter, which is much computationally fast [19]. Another advantage of this approximation is that convolution with box filter can be easily calculated with the help of integral images and can be performed in parallel for different scales [3]. For these reasons, we detect SURF interest points as keypoints for our scheme. Note that we are only using the SURF keypoint detectors to find the high entropy SURF keypoints in the image with comparatively lesser computational complexity.

**Fig. 5.2** SURF keypoints with orientation

### *5.2.2 RLBP Feature Extraction*

After finding the SURF keypoints, we extract the *Rotated Local Binary Pattern* (RLBP) features corresponding to each SURF keypoints in a circular neighborhood. RLBP is a rotation-invariant version of the popular *Local Binary Pattern* (LBP) [16] texture feature. LBP is calculated in a local (circular region) neighborhood by taking the difference of the center pixel with respect to its neighboring gray scale pixel values as follows:

$$LBP_{R,P} = \sum_{k=0}^{P-1} s(g_k - g_c) \cdot 2^k \tag{5.1}$$

$$\text{where} \quad s(g_k - g_c) = \begin{cases} 1, & \text{for } g_k \geq g_c \\ 0, & \text{for } g_k \leq g_c \end{cases} \tag{5.2}$$

Here, $g_c$ and $g_k$ denote the gray scale values (intensity) of the center pixel and its neighbors, respectively, $R$ is the radius of the circular neighborhood, $k$ denotes the index of the neighbor, $P$ denotes the number of neighboring pixels, and $s(g_k - g_c)$ denotes the weights of $g_k$ with respect to the center pixel.

Local Binary Pattern (LBP) is not rotation-invariant. Hence, to overcome this limitation, recently, a modified rotation-invariant texture descriptor, viz., *Rotated Local Binary Pattern* (RLBP) using dominant direction [16] has been introduced. The *dominant direction* ($D$) in a neighborhood is indicated by the index of the neighbor whose difference to the center pixel is maximum, i.e.,

$$D = \max_{k \in (0,1,\ldots,P-1)} |g_k - g_c| \quad (5.3)$$

If we consider that there are not multiple dominant directions, then the dominant direction would be rotation-invariant. Therefore, any circular shift of the weights with respect to the dominant direction results rotation-invariant RLBP features, which can be defined as follows:

$$RLBP_{R,P} = \sum_{k=0}^{P-1} s(g_k - g_c) \cdot 2^{(k-D)\,(\mathrm{mod}\,P)} \quad (5.4)$$

For each detected SURF keypoints, we choose a 21×21 neighborhood and compute the $RLBP_{4,8}$ features for each of the keypoint. Then the RLBP histogram is taken as features; therefore, feature dimension is 256.

The reason behind selecting RLBP features is the fact that the tampered image must contain some edge abnormality-induced artifacts which have been introduced during the copy-move tampering operation. Hence, the tampered image seems to be more textured than the pristine image since it subsumes some local artifacts at the boundary regions of the forgery. Rotated local binary pattern, which is an improved texture feature along with its additional rotational invariance capability, is thus able to efficiently distinguish between the pristine and tampered image.

#### 5.2.2.1 Rotation Invariance of RLBP

The major drawback of LBP is that it is not rotation-invariant, which is overcome in our case taking rotation-invariant RLBP. As shown in Eq. (5.4), if we have a rotated version of the neighborhood, then also the dominant direction would be the same as defined in Eq. (5.3) and the weights are circularly shifted with respect to the dominant direction as the weighted term $2^{mod(p-D,P)}$ depends on the dominant direction ($D$). Hence, RLBP features are effective for geometric transformation-based copy-move image forgery.

#### 5.2.2.2 Illumination Invariance of RLBP

It has been observed that with respect to illumination change operation, the CMFD performance of the state-of-the-art schemes [2, 6, 26] degrade significantly particularly in the case of images containing similar but genuine objects. Hence, illumination-invariant texture features, e.g., LBP, would be effective in such cases [16]. Particularly, RLBP is found to be more resistant to illumination alteration than standard LBP since it is computed based on the difference of the pixel intensity values with respect to the dominant direction, which is rotation-invariant. Experimental validation in Sect. 5.3 verifies this for CMFD.

We provide a brief justification of this quality of RLBP in the present context as follows. If we look into image pixel intensity following the *Lambertian model* [27], we can consider [17],

$$I_c = \rho_c\, \mathbf{n_c^T} \cdot \mathbf{s_c} \qquad I_p = \rho_p\, \mathbf{n_p^T} \cdot \mathbf{s_p} \tag{5.5}$$

where $\rho$ is the albedo, $\mathbf{n^T}$ is the surface normal, $\mathbf{s}$ is the incident illumination, $I_c$ and $I_p$ are the center pixel and neighbor pixel intensities, respectively. Although $\rho$ and $n^T$ depend on surface property of the object, still the parameters are expected to be similar for an image in a small spatial neighborhood, which has been considered in case of RLBP. Hence, difference of pixel intensities described by the RLBP operator actually represents the change in illumination level, i.e.,

$$I_p - I_c = \rho\, \mathbf{n^T} \cdot \left(\mathbf{s_p} - \mathbf{s_c}\right) \tag{5.6}$$

Hence, if we consider the illumination change to be monotonic, then the features have invariant behavior as evident from Eqs. (5.4) to (5.6), assuming $\rho$ and $\mathbf{n^T}$ to be constant for both pixels.

### 5.2.3 Feature Matching

After RLBP feature extraction, the next step is to perform feature matching. Here, we use the *generalized 2-Nearest Neighborhood* (g2NN) feature matching technique [2] to identify the possible matches among the entire feature space.

Initially, the best candidate match for each keypoint is chosen with minimum Euclidean distance among all other keypoints in the feature space, i.e., the nearest neighbor keypoint. However, improvement of this has been performed using 2-Nearest Neighborhood (2NN) test. This test compares whether the ratio between the distance of the candidate match (i.e., the nearest neighbor keypoint) and distance of the second nearest neighbor keypoint is lower than a predefined threshold $T$. For example, given a keypoint, suppose the vector $D = \{d_1, d_2, \ldots, d_n\}$ represents sorted Euclidean distances with respect to other descriptors. Then, the keypoint is matched if $d_1/d_2 \leq T$.

The g2NN generalizes the feature matching considering iterative 2-Nearest Neighborhood (2NN) test between $d_i/d_{i+1}$, until the ratio becomes greater than a predefined threshold value ($T$). The corresponding matches for the keypoint will be the keypoints with distance in $\{d_1, d_2, \ldots, d_k\}$, where $k$ is the value where the process stops. In case of our experiments, the threshold value has been taken as, $T = 0.5$.

### 5.2.4 Clustering and Forgery Detection

Feature matching finally produces one to one correspondence among the set of matched keypoints. Next, we have to identify the possible similar areas by an *agglomerative hierarchical clustering* [9]. This clustering is performed on the spatial location of the matched feature points with *Ward's linkage* [2]. Such clustering typically generates tree-structured hierarchy of clusters. At the outset, each keypoint is allotted to a cluster, and then all the reciprocal spatial distances among the clusters are computed and based on that the proximate pair of clusters is determined and eventually merges them to constitute a single cluster. The same process is iterated until the linkage-based final merging happens.

As an instance, let us assume that the two clusters, viz., $A$ and $B$ contains $n_A$, $n_B$ objects, respectively, and $x_{Ai}$, $x_{Bj}$ denote the $i$th and $j$th object in the clusters, respectively. Then, *Ward's linkage* estimates the increase or decrease in the Error Sum of Squares (ESS) after merging the two clusters into a single cluster with respect to the case of two distinct clusters:

$$d_{dist}(A, B) = ESS(AB) - [ESS(A) + ESS(B)] \tag{5.7}$$

where $\overline{x_A}$ denotes the arithmetic mean of the cluster $A$, and,

$$ESS(A) = \sum_{i=1}^{n_A} |x_{Ai} - \overline{x_A}|^2 \tag{5.8}$$

The spatially close keypoints are grouped together by this clustering process. Finally, we eliminate those clusters having less than three keypoints since they would not contribute to any geometric transformation estimation.

Once clustering is done, we try to estimate the geometric transformations, if any, among the clusters. Generally, minimum three matched point pairs between each cluster are required to calculate an affine geometric transformation for estimating the geometrical transformation between matched points among the clusters. We also proceed with this approach.

Suppose, the coordinates of the matched points be $\{(\mathbf{x_1}, \mathbf{x_1'}), \ldots, (\mathbf{x_k}, \mathbf{x_k'})\}$, where keypoint coordinates are denoted as $\mathbf{x_i} = (x_i, y_i)$. Here, keypoint $\mathbf{x}$ resides in one cluster and keypoint $\mathbf{x'}$ resides in another cluster. Then the affine transformation can be expressed as

$$\begin{pmatrix} x' \\ y' \\ 1 \end{pmatrix} = \begin{pmatrix} a_{11} & a_{12} & t_x \\ a_{21} & a_{22} & t_y \\ 0 & 0 & 1 \end{pmatrix} \begin{pmatrix} x \\ y \\ 1 \end{pmatrix} = \mathbf{H} \begin{pmatrix} x \\ y \\ 1 \end{pmatrix} \tag{5.9}$$

where $a_{11}, a_{12}, a_{21}, a_{22}$ denote the scaling and rotation direction deformation; $t_y, t_x$ denote the translational counterpart; $\mathbf{H}$ denotes the affine homography which is estimated by *normalized Direct Linear Transform* (DLT) [11]. The false matches are eliminated by *RANdom SAmple Consensus* (RANSAC) algorithm [11]. For each cluster pair (corresponding to pristine and duplicated region) an affine geometric transformation is computed [1]. Next, based on the count of the estimated affine transform, we decide the image to be forged or not. The image under inspection is considered to be forged if the corresponding count of the total estimated affine transformation is nonzero. If no region similarity through affine transformation is found, then the image is considered to be pristine. Therefore, the number of estimated affine transform has been considered as a measure of forgery detection.

The main idea is that tampered image contains similar regions which can be related by a geometric transformation, which is unlikely in the case of an original image.

## 5.3 Experimental Results

### 5.3.1 Experimental Setup

Our forgery detection approach has been evaluated and compared with the state-of-the-art SIFT [2], SURF [26], Dense-Field [6], SHFD [28] based methods, on the recently proposed COVERAGE database which contains 100 pairs of pristine and copy-move forged image containing SGO. In the database, six types of different image tampering, viz., translation, rotation, scaling, illumination, free-form, and combined factors have been considered. We have also verified our technique on other benchmark databases, viz., *CoMoFoD* [24] and *Manipulate* [5].

For performance evaluation of CMFD, we have used detection accuracy ($Acc$) as a metric:

$$Acc = \frac{TP + TN}{TP + TN + FN + FP} \times 100(\%) \tag{5.10}$$

where $TN$, $TP$, $FN$, and $FP$ denote true negative (correctly predicted authentic), true positive (correctly predicted tampered), false negative (tampered predicted as authentic), and false positive (authentic predicted as tampered), respectively.

### 5.3.2 Comparison with State of the Art

Comparison of the proposed approach with several state-of-the-art techniques (with different attacks) has been performed in Table 5.1 for COVERAGE dataset. Table 5.1 clearly shows that the proposed technique performs best with respect to the state of the art for COVERAGE dataset. It also performs consistently well with respect to other datasets as shown in Table 5.2 when compared to the state of the art. Figures 5.3 and 5.4 exemplify some of the forgery detection instances.

**Table 5.1** CMFD detection accuracy (%) comparison for COVERAGE with the state of the art

| Operation (# test image) | SIFT [2] | SURF [26] | SHFD [28] | Dense field [6] | Proposed scheme |
|---|---|---|---|---|---|
| Translation (16) | 50.0 | 75.0 | 75.0 | **93.8** | **93.8** |
| Scale (16) | 56.3 | 56.3 | 50.0 | **75.0** | 50.0 |
| Rotation (16) | 46.7 | 53.3 | 56.2 | **86.7** | 62.5 |
| Free-form (16) | 43.8 | 50.0 | 45.0 | 68.8 | **75.0** |
| Illumination (16) | 43.3 | 62.5 | 45.0 | 62.5 | **81.3** |
| Combined (20) | **55.0** | **55.0** | 43.7 | 50.0 | 45.0 |
| Original (100) | 71.8 | 52.3 | 64.6 | 60.2 | **73.0** |
| Total (200) | 60.5 | 55.5 | 58.5 | 66.5 | **70.5** |

**(a) SIFT** **(b) SURF** **(c) SHFD** **(d) RLBP**

**Fig. 5.3** CMFD performance comparison of SIFT [2], SURF [26], SHFD [28], and proposed RLBP-based schemes (illumination tampering) on COVERAGE image database. The upper row contains original images with SGO, and the lower row contains forged image. Dots and lines denote keypoints and matched pairs, respectively

**Table 5.2** Detection accuracy comparison for other databases

| CMFD database (# test image) | SIFT | SURF | SHFD | Proposed scheme |
|---|---|---|---|---|
| CoMoFoD (200) | **77.0** | 51.5 | 27.0 | 70.1 |
| Manipulate (48) | 75.0 | 58.3 | 20.8 | **83.3** |

**(a) Original** **(b) Forged**

**Fig. 5.4** Comparison of the proposed scheme in different databases: COVERAGE (first row), Manipulate (second row), and CoMoFoD (third row)

Table 5.1 also reveals that the proposed scheme outperforms the existing schemes with respect to free-form, illumination-based attacks on the COVERAGE dataset. The proposed scheme performs comparably with the state of the art for other attacks also. The best CMFD results are marked in bold. In all our experiments, we have taken

$RLBP_{4,8}$ with feature radius 4, since this produces the best experimental results with respect to CMFD accuracy as shown in Fig. 5.5.

### 5.3.3 *Experiments on Post-processed Tampered Images*

The forgery detection methodology should be robust against different post-processing operations. We have validated the robustness of our scheme with respect to different post-processing operations applied on the tampered images, viz., filtering, compression, blurring, etc. We have post-processed the images from the COVERAGE dataset using Gaussian blurring (window size $w = 3$ and sigma $\sigma \in \{0.5, 1\}$) and JPEG compression (with quality factors $Q \in \{60, 80\}$). Experimental results in Table 5.3 verify that our proposed scheme is robust to these post-processing operations with respect to the state-of-the-art techniques on the COVERAGE dataset.

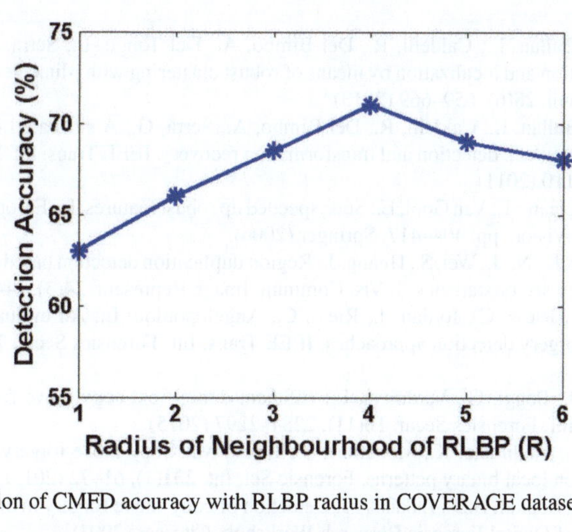


**Fig. 5.5** Variation of CMFD accuracy with RLBP radius in COVERAGE dataset

**Table 5.3** Detection accuracy after post-processing for COVERAGE

| Post-processing operation | SIFT | SURF | Dense field | SHFD | Proposed approach |
|---|---|---|---|---|---|
| Compression (Q = 80) | 56 | 53.5 | 66.5 | 57.5 | **70.5** |
| Compression (Q = 60) | 54 | 54 | **66.5** | 42 | 65.5 |
| Blurring ($\sigma = 0.5$) | 60.5 | 55 | 66 | 56.5 | **68** |
| Blurring ($\sigma = 1$) | 60 | 55.5 | **65** | 54 | 64.5 |

## 5.4 Conclusion

Copy-move forgery detection with similar but genuine objects has been found to be more challenging and less explored with respect to the existing CMFD techniques in the current literature. Such techniques treat region similarity as a measure of forgery detection, therefore, undergoes performance degradation due to false positives in this particular case. In this chapter, an efficient CMFD with SGO has been proposed using RLBP features along with g2NN-based feature matching, hierarchical agglomerative clustering, and geometric transformation estimation. This technique performs best for CMFD for SGO in the COVERAGE dataset and also consistently well for other benchmark datasets for CMFD with respect to the state-of-the-art techniques. The scheme has also found to be robust with respect to post-processing of the forgery.

## References

1. Amerini, I., Ballan, L., Caldelli, R., Del Bimbo, A., Del Tongo, L., Serra, G.: Copy-move forgery detection and localization by means of robust clustering with j-linkage. Signal Process. Image Commun. **28**(6), 659–669 (2013)
2. Amerini, I., Ballan, L., Caldelli, R., Del Bimbo, A., Serra, G.: A sift-based forensic method for copy-move attack detection and transformation recovery. IEEE Trans. Inf. Forensics Secur. **6**(3), 1099–1110 (2011)
3. Bay, H., Tuytelaars, T., Van Gool, L.: Surf: speeded up robust features. In: European Conference on Computer Vision, pp. 404–417. Springer (2006)
4. Chen, L., Wei, L., Ni, J., Wei, S., Huang, J.: Region duplication detection based on Harris corner points and step sector statistics. J. Vis. Commun. Image Represent. **24**(3), 244–254 (2013)
5. Christlein, V., Riess, C., Jordan, J., Riess, C., Angelopoulou, E.: An evaluation of popular copy-move forgery detection approaches. IEEE Trans. Inf. Forensics Secur. **7**(6), 1841–1854 (2012)
6. Cozzolino, D., Poggi, G., Verdoliva, L.: Efficient dense-field copy-move forgery detection. IEEE Trans. Inf. Forensics Secur. **10**(11), 2284–2297 (2015)
7. Davarzani, R., Yaghmaie, K., Mozaffari, S., Tapak, M.: Copy-move forgery detection using multiresolution local binary patterns. Forensic Sci. Int. **231**(1), 61–72 (2013)
8. Fridrich, A.J., Soukal, B.D., Lukáš, A.J. Detection of copy-move forgery in digital images. In: Proceedings of Digital Forensic Research Workshop. Citeseer (2003)
9. Friedman, J., Hastie, T., Tibshirani, R.: The Elements of Statistical Learning, vol 1. Springer series in statistics Springer, Berlin (2001)
10. Guo, J., Liu, Y.-F., Wu, Z.-J.: Duplication forgery detection using improved daisy descriptor. Expert. Syst. Appl. **40**(2), 707–714 (2013)
11. Hartley R., Zisserman A.: Multiple View Geometry in Computer Vision. Cambridge University Press, Cambridge (2003)
12. Huang, H., Guo, W., Zhang, Yu.: Detection of copy-move forgery in digital images using sift algorithm. In: Pacific-Asia Workshop on Computational Intelligence and Industrial Application, PACIIA'08, vol. 2, pp. 272–276. IEEE (2008)
13. Kang, X., Wei, S.: Identifying tampered regions using singular value decomposition in digital image forensics. In: 2008 International Conference on Computer Science and Software Engineering, vol. 3, pp. 926–930. IEEE (2008)
14. Li, W., Yu, N.: Rotation robust detection of copy-move forgery. In: 2010 17th IEEE International Conference on Image Processing (ICIP), pp. 2113–2116. IEEE (2010)

15. Lowe, D.G.: Distinctive image features from scale-invariant keypoints. Int. J. Comput. Vis. **60**(2), 91–110 (2004)
16. Mehta, R., Egiazarian, K.O.: Rotated local binary pattern (rlbp)-rotation invariant texture descriptor. In: ICPRAM, pp. 497–502 (2013)
17. Méndez-Vázquez, H., García-Reyes, E., Condes-Molleda, Y.: A new combination of local appearance based methods for face recognition under varying lighting conditions. In: Iberoamerican Congress on Pattern Recognition, pp. 535–542. Springer (2008)
18. Muhammad, G., Muhammad, H., Bebis, G.: Passive copy move image forgery detection using undecimated dyadic wavelet transform. Digit. Investig. **9**(1), 49–57 (2012)
19. OpenCV Documentation. Introduction to surf. https://docs.opencv.org/3.0-beta/doc/py_tutorials/py_feature2d/py_surf_intro/py_surf_intro.html#additional-resources (2013)
20. Pan, X., Lyu, S.: Region duplication detection using image feature matching. IEEE Trans. Inf. Forensics Secur. **5**(4), 857–867 (2010)
21. Popescu, A.C., Farid, H.: Exposing digital forgeries by detecting duplicated image regions. Technical Report TR2004-515, Department of Computer Science, Dartmouth College (2004)
22. Roy, A., Konda, A., Chakraborty, R.S.: Copy move forgery detection with similar but genuine objects. In: 2017 IEEE International Conference on Image Processing (ICIP), pp. 4083–4087. IEEE (2017)
23. Ryu, S.-J., Lee, M.-J. Lee, H.-K.: Detection of copy-rotate-move forgery using zernike moments. In: International Workshop on Information Hiding, pp. 51–65. Springer (2010)
24. Tralic, D., Zupancic, I., Grgic, S., Grgic, M.: Comofod–new database for copy-move forgery detection. In: 2013 55th International Symposium on ELMAR, pp. 49–54. IEEE (2013)
25. Wen, B., Zhu, Y., Subramanian, R., Ng, T.-T., Shen, X., Winkler, S.: Coverage–a novel database for copy-move forgery detection. In 2016 IEEE International Conference on Image Processing (ICIP), pp. 161–165. IEEE (2016)
26. Xu, B., Wang, J., Liu, G., and Dai, Y.: Image copy-move forgery detection based on surf. In: 2010 International Conference on Multimedia Information Networking and Security (MINES), pp. 889–892. IEEE (2010)
27. Yuille, A.: Lecture 10: the lambertian reflectance model (2012)
28. Zhu, Y., Ng, T.-T., Shen, X., Wen, B.: Revisiting copy-move forgery detection by considering realistic image with similar but genuine objects. arXiv preprint. arXiv:1601.07262 (2016)

# Chapter 6
# Copy-Move Forgery Detection in Transform Domain

## 6.1 Introduction

During the past decade, a number of significant research works have been carried out toward identifying region duplication or copy-move forgery in digital images. Important works in this direction include [2, 4, 5, 10], which all perform image duplication identification through blind digital forensics. In this chapter, we present a region duplication detection technique, utilizing Dyadic Wavelet Transform (DyWT) [3]. In the presented method, we divide a given image into overlapping blocks and extract the features (energy coefficient of LL sub-band) using DyWT, subsequently performing block matching via calculated using Canberra distance matrices. Subsequently, we present a eight-neighbor block checking method to optimize the number of false matches.

The rest of the chapter is organized as follows. The proposed technique is presented in Sect. 6.2 in detail. Experimental results are presented and discussed in Sect. 6.3. The chapter summary has been presented in Sect. 6.4.

## 6.2 DyWT-Based Image Region Duplication Detection

In this technique, to identify the duplicated image regions, initially the image is transformed into low- and high-frequency components using undecimated Dyadic Wavelet Transform (DyWT) [3]. Further, these frequency components are divided into fixed size overlapping blocks. Here, Canberra distance [1] metric is used to find the similarities between individual block pairs, based on which the blocks are sorted lexicographically. If the similarities exceed a certain predefined threshold, then the concerned blocks are detected to be duplicated. In the method presented next, the low-frequency components are used to measure the block similarities.



A. Roy et al., *Digital Image Forensics*, Studies in Computational Intelligence 755, https://doi.org/10.1007/978-981-10-7644-2_6

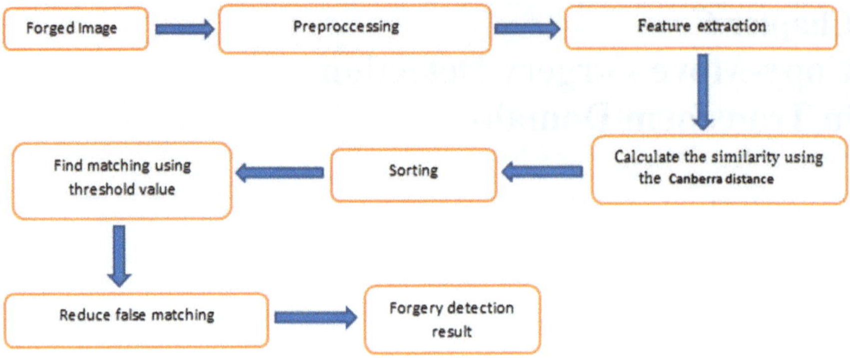


**Fig. 6.1** Operational flowchart

The method consists of five basic steps:

1. Pre-processing of the image so that the image is ready for feature extraction;
2. Feature extraction;
3. Evaluating similarity among individual image blocks;
4. Sorting the image blocks based on their similarities; and
5. Detection of duplicate blocks from the above list.

The rate of false block matching or *false positives* is reduced by a neighborhood check technique, applied on similar block pairs. A block diagram representing the operation of the proposed method is shown in Fig. 6.1. The possibly forged color image is pre-processed by converting it into the gray scale version as follows:

$$I = 0.299 \times R + 0.587 \times G + 0.114 \times B \tag{6.1}$$

where $R$, $G$, and $B$ represent the red, green, and blue color channel intensities of the image, and $I$ represents its gray scale version. This gray scale image is used for feature extraction. By applying DyWT on the gray scale image, the image is decomposed into the four frequency sub-bands: $LL$ (approximation), $LH$ (horizontal), $HL$ (vertical), and $HH$ (diagonal) sub-bands. Figure 6.2 depicts one-level decomposition of an image using DyWT. At scale one, $LL_1$ and $HH_1$ sub-bands provide information about similarity and dissimilarity between copied and moved image blocks, respectively.

The similarities between the individual image blocks are measured by using the *Canberra distance* here. Initially, the $LL_1$ sub-band is divided into fixed size overlapping blocks. Let us assume that the size of the image and that of each block is $w \times h$ and $b \times b$ pixels, respectively. Hence, the total number of blocks is given by: $(w - b + 1) \times (h - b + 1)$. Next, the similarity between the individual block pairs is measured using the Canberra distance metric [1], as follows:

**Fig. 6.2** DyWT decomposition: **a** Test image, and, **b** $LL_1$ sub-band **c** $HH_1$ sub-band

$$D(x, y) = \sum_{i=1}^{n} \frac{|x_i - y_i|}{|x_i| - |y_i|} \tag{6.2}$$

Here, $D(x, y)$ indicates the Canberra distance between blocks $x$ and $y$, respectively, and $n$ represents the total number of pixels in a block. The value of Canberra distance lies in the range [0, 1]. In the same way, the $HH_1$ sub-band is next is divided into overlapping blocks, and subsequently, we calculate the similarity or Canberra distance between those blocks. The (similar) $LL_1$ block pairs are arranged orderly in $List_1$. Similarly, the (dissimilar) $HH_1$ blocks are arranged in descending order in $List_2$. Hence in $List_1$ and $List_2$, highly similar pairs of blocks and highly dissimilar pairs of blocks are arranged, respectively.

Block matching is performed using similarities and dissimilarities between block pairs. For this purpose, we need two threshold parameters, i.e., $TH_1$ (threshold value for $List_1$) and $TH_2$ (threshold value for $List_2$). The threshold parameters are selected empirically, through experiments. Block pairs $(x, y)$ from $List_1$, for which distance $D(x, y) < TH_1$ are considered, and arranged in ascending order. Similarly, block pairs $(x, y)$ from $List_2$, for which $D(x, y) < TH_2$ are considered and arranged in descending order. In our experiments, we select the value of $TH_1$ and $TH_2$ as 0.8 and 0.2, respectively. The pairs of blocks fulfilling the threshold criteria represent the duplicate regions in an image.

Since block matching in the presented method is based on finding similarities from $LL_1$ sub-band, this leads to false positives. As $HH_1$ analyzes the dissimilarity in noise and edge patterns, it is not possible to use those in detection of copy-move forged regions in an image. There also exist some false positives due to $HH_1$ as it focuses on high-frequency information. On the other hand, $LL_1$ is expected to be better in finding find the duplicate regions as it is obtained by applying low-pass filter and focuses on low-frequency features.

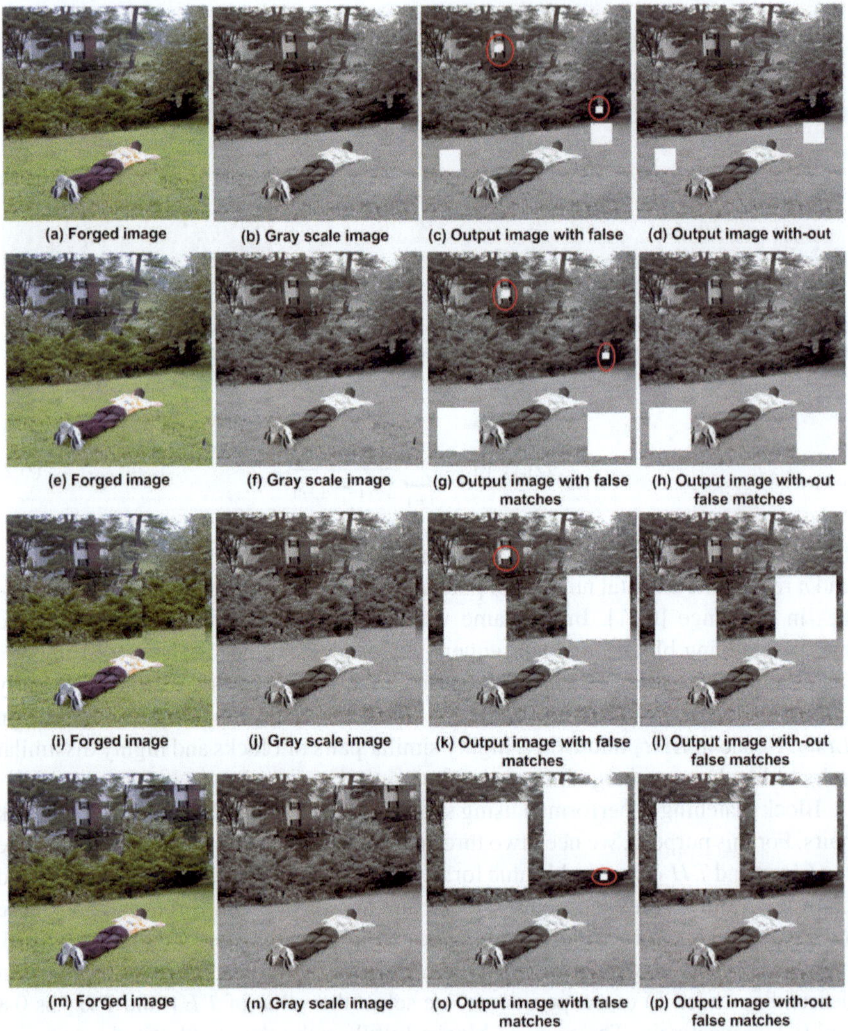


**Fig. 6.3** Detection of copy-move forgery (of varying sizes), before and after reduction of false matches. Forgery sizes: **a** 10%; **d** 20%; **g** 30%, and, **j** 40%. Detected duplicate regions are highlighted in white. False positives are shown with red circles

## 6.2.1 Minimization of False Matches

We select block pairs $P$ and $Q$ from $List_1$ and $List_2$, respectively, and find out their eight neighbors (two neighbors in horizontal, two in vertical, and four in diagonal directions) of each of $P$ and $Q$ blocks. Let $N_{nbr-test}$ denote the number of eight neighbors of a block pair. Block pairs for which $N_{nbr-test} \geq 4$ (for both blocks minimum four neighbors are same), are considered to be duplicated; otherwise, we

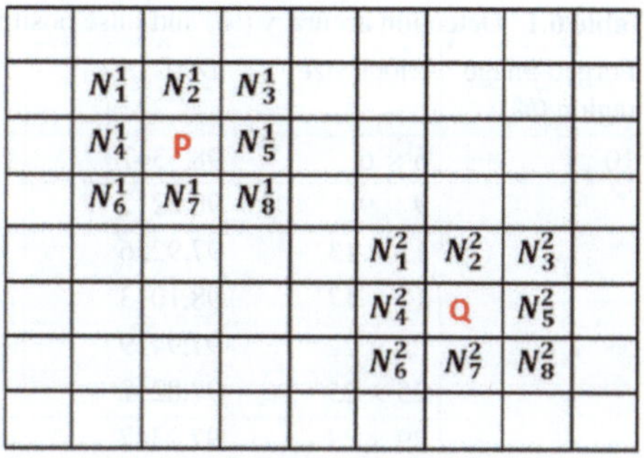


**Fig. 6.4** Selected neighboring blocks of $P$ and $Q$

discard $P$ and $Q$, and move on to the next pair. For example, in Fig. 6.4, $N_1^1, \ldots, N_8^1$ are the neighbors of $P$ and $N_1^2, \ldots, N_8^2$ are the neighbors of $Q$. At least four out of $N_1^1, \ldots, N_8^1$, as well as four out of $N_1^2, \ldots, N_8^2$ must be present in both lists. Only then, $P$ and $Q$ are considered to be duplicate blocks. The detection of copy-move forgery, and the output after reducing false positives using eight-neighbor check presented above, is shown in Fig. 6.3. In Fig. 6.3, all forged images (size of forgery varied from 10 to 40%) have been shown in the left and corresponding false matches in the middle. Corresponding images, after reducing false positives are shown in the right.

## 6.3 Experimental Results

In this section, we analyze the performance of the presented method and discuss our experimental results. We use a dataset of 60 standard test images of size $512 \times 512$, taken from the USC SIPI [9] and CVG UGR [8] image databases. The forged region size is varied to be 10, 20, 30, and 40% of the given image. The block size is varied from $6 \times 6$ to $34 \times 34$. Our experimental results are presented in Table 6.1. $DAC_1$ and $FPR_1$ indicate the initial detection accuracy and false positive rate, while $DAC_2$ and $FPR_2$ are the detection accuracy and false positive rate after reducing false matches using eight-neighbor check (discussed in Sect. 6.2.1). According to Table 6.1, initially the detection accuracy and false positive rates were between 96.97% to 99.21% and 3.04% to 15.21%, respectively. After application of the eight-neighbor check, the detection accuracy and false positive rates improved to 99.73% to 99.78% and 1.20% to 2.06%, respectively.

Figure 6.5 shows the variation of detection accuracy and false positive rate, with block size as well as the percentage of image forgery, both before and after false positives reduction. Table 6.2 presents the results of the comparison of the proposed method with three recent existing region duplication detection algorithms [5–7]. From Table 6.2, it is clear that detection accuracy of the method presented in this chapter is higher with false positive rate as low as 1.2–2.3%.

**Table 6.1** Detection accuracy (%) and false positive rate (%) results

| Forged image region (%) | Block size | $DAC_1$ (%) | $FPR_1$ (%) | $DAC_2$ (%) | $FPR_2$ (%) |
|---|---|---|---|---|---|
| 10 | 6 × 6 | 98.5347 | 8.1782 | 99.7378 | 2.0382 |
| | 9 × 9 | 98.2272 | 8.6724 | 99.7304 | 2.0612 |
| | 13 × 13 | 97.9256 | 9.8495 | 99.7389 | 2.0387 |
| | 17 × 17 | 98.1013 | 8.8218 | 99.7398 | 1.9974 |
| | 22 × 22 | 97.9419 | 10.4917 | 99.7411 | 1.9914 |
| | 25 × 25 | 97.8212 | 12.5370 | 99.7501 | 1.9462 |
| | 29 × 29 | 97.2347 | 14.5372 | 99.7588 | 1.9272 |
| | 34 × 34 | 96.9732 | 15.2155 | 99.7601 | 1.9012 |
| 20 | 6 × 6 | 98.8472 | 6.8797 | 99.7398 | 2.0112 |
| | 9 × 9 | 98.1173 | 7.8780 | 99.7362 | 1.8122 |
| | 13 × 13 | 98.9878 | 6.7346 | 99.7382 | 1.8083 |
| | 17 × 17 | 97.9757 | 7.2302 | 99.7422 | 1.8013 |
| | 22 × 22 | 97.9850 | 9.8669 | 99.7511 | 1.7619 |
| | 25 × 25 | 97.8218 | 12.7585 | 99.7528 | 1.7111 |
| | 29 × 29 | 97.9712 | 10.9244 | 99.7616 | 1.6732 |
| | 34 × 34 | 97.8493 | 13.6799 | 99.7724 | 1.6317 |
| 30 | 6 × 6 | 98.9053 | 6.8807 | 99.7421 | 1.9081 |
| | 9 × 9 | 98.2174 | 5.6808 | 99.7460 | 1.6085 |
| | 13 × 13 | 98.9076 | 6.7295 | 99.7503 | 1.3938 |
| | 17 × 17 | 98.7368 | 6.9202 | 99.7530 | 1.3920 |
| | 22 × 22 | 98.4889 | 7.8969 | 99.7620 | 1.3237 |
| | 25 × 25 | 98.4038 | 8.7085 | 99.7698 | 1.3482 |
| | 29 × 29 | 98.3410 | 8.9746 | 99.7706 | 1.3012 |
| | 34 × 34 | 98.5143 | 8.2689 | 99.7788 | 1.2511 |
| 40 | 6 × 6 | 99.2173 | 3.7807 | 99.7429 | 1.4081 |
| | 9 × 9 | 98.9245 | 4.9846 | 99.7483 | 1.3821 |
| | 13 × 13 | 98.8901 | 5.1560 | 99.7513 | 1.3748 |
| | 17 × 17 | 99.3564 | 3.0402 | 99.7580 | 1.3522 |
| | 22 × 22 | 99.1433 | 4.1689 | 99.7680 | 1.3018 |
| | 25 × 25 | 98.8798 | 6.5871 | 99.7737 | 1.2882 |
| | 29 × 29 | 98.8354 | 6.1031 | 99.7787 | 1.2191 |
| | 34 × 34 | 98.9823 | 5.6921 | 99.7827 | 1.2024 |

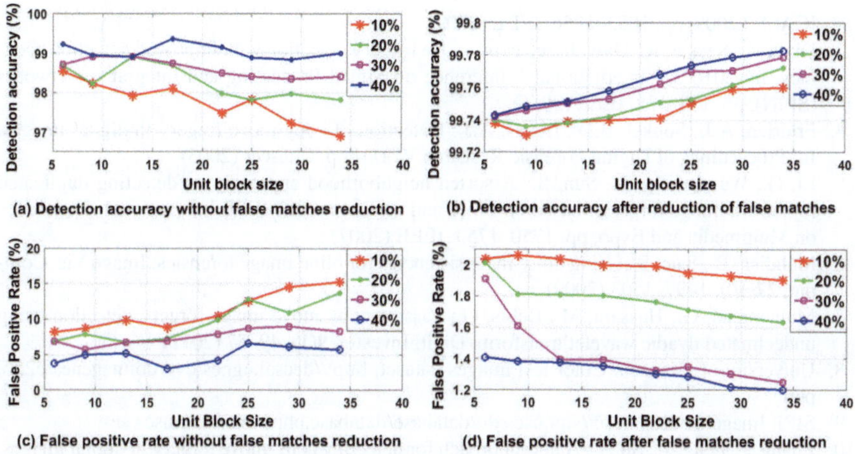


**Fig. 6.5** Variation of detection accuracy and false positive rates: **a** detection accuracy versus unit block size with false matches; **b** detection accuracy versus unit block size after false positive reduction; **c** false positive rate versus unit block size with false matches; and **d** false positive rate versus unit block size after false positive reduction

**Table 6.2** Comparison with existing techniques

| Method | DA (%) | FPR (%) |
|---|---|---|
| DyWT-based [7] | 95.90 | 4.54 |
| SVD-based [5] | 91.03 | 9.65 |
| Noise inconsistencies-based [6] | 81.18 | 10.03 |
| Proposed method | 99.73–99.78[a] | 1.2–2.3[a] |

[a]Varies with block size

## 6.4 Conclusion

In this chapter, a copy-move forgery detection method based on DyWT image decomposition has been presented. We also present a false positives minimization technique based on eight-neighborhood check. The experimental results suggest the superiority of the above technique as compared to the state-of-the-art techniques.

## References

1. Androutsos, D., Plataniotiss, K.N., Venetsanopoulos, A.N.: Distance measures for color image retrieval. In: Proceedings 1998 International Conference on Image Processing, 1998. ICIP 98, vol. 2, pp. 770–774. IEEE (1998)
2. Bayram, S., Sencar, H.T., Memon, N.: An efficient and robust method for detecting copy-move forgery. In: IEEE International Conference on Acoustics, Speech and Signal Processing, 2009.

ICASSP 2009, pp. 1053–1056. IEEE (2009)
3. Dixit, R., Naskar, R.: Dywt based copy-move forgery detection with improved detection accuracy. In: 2016 3rd International Conference on Signal Processing and Integrated Networks (SPIN), pp. 133–138. IEEE (2016)
4. Fridrich, A.J., Soukal, B.D., Lukáš, A.J.: Detection of copy-move forgery in digital images. In: Proceedings of Digital Forensic Research Workshop. Citeseer (2003)
5. Li, G., Wu, Q., Tu, D., Sun, S.: A sorted neighborhood approach for detecting duplicated regions in image forgeries based on DWT and SVD. In: 2007 IEEE International Conference on Multimedia and Expo, pp. 1750–1753. IEEE (2007)
6. Mahdian, B., Saic, S.: Using noise inconsistencies for blind image forensics. Image Vis. Comput. **27**(10), 1497–1503 (2009)
7. Muhammad, G., Hussain, M., Bebis, G.: Passive copy move image forgery detection using undecimated dyadic wavelet transform. Digit. Investig. **9**(1), 49–57 (2012)
8. University of Granada: Color test images dataset. http://decsai.ugr.es/cvg/dbimagenes/c256.php
9. SIPI: Image dataset. http://sipi.usc.edu/database/database.php?volume=misc
10. Zhang, J., Feng, Z., Su, Y.: A new approach for detecting copy-move forgery in digital images. In: 11th IEEE Singapore International Conference on Communication Systems, 2008. ICCS 2008, pp. 362–366. IEEE (2008)

# Chapter 7
# Conclusion and Future Research Directions

Digital images are one of the primary modern media for information interchange. However, digital images are vulnerable to interception and manipulation due to user-friendly, widely available and less expensive (often free of cost) image editing tools. Therefore, blind authenticity and fidelity verification of such digital media are of utmost importance nowadays. Hence, *digital image forensics* has become an important field of research of late. In this book, we have presented in detail two important problems of digital image forensics, viz., camera source identification and image forgery detection. Starting with the camera source identification problem, we gradually moved to a topic of great current research interest, viz., *copy-move forgery detection*. Following a detailed survey and discussion of this type of forgery detection, we described statistical feature analysis based automated copy-move forgery detection. We then described an interesting and challenging type of copy-move forgery detection in images with similar but genuine objects.

Since digital image forensics is a budding research domain, there are substantial open problems that merit future research:

- We have mainly covered copy-move forgery in this book. Therefore, other forms of image forgery, e.g., *splicing attacks*, image filtering, distinguishing natural and computer-generated image, etc., are problems that can be investigated further.
- An extension of copy-move forgery detection is the identification of geometrically transformed duplicated image regions. Geometric transformations may include rescaling, rotation, reflection or a combination of two or more of these.
- Copy-move forgery detection with similar but genuine objects is relatively unexplored, and performance of the scheme proposed in this book and newer schemes need to be explored for different image datasets.
- Existing state-of-the-art digital image forensic techniques utilize several image texture and computer vision inspired features to train a classifier using machine learning based model building and prediction. However, it is observed that different features are effective for different types of image forgery, e.g., computer



A. Roy et al., *Digital Image Forensics*, Studies in Computational Intelligence 755, https://doi.org/10.1007/978-981-10-7644-2_7

vision inspired SIFT or SURF features are effective for copy-move, geometric transformation-based forgery attacks, whether for detecting recompression or splicing attacks DCT domain statistical features are more effective. In practical scenario, the forensic analyst must be unaware of the type of forgery the image had undergone; hence, it is motivating to come up with a universal [1, 4] as well as efficient image manipulation detection technique that is capable of detecting any class of image forgery. Recently, deep learning breakthroughs have boosted every aspect of machine learning research and its applications, including digital forensics. The major limitation of the state-of-the-art image forensic techniques is that those rely on the identification of efficient image features for individual types of forgeries. This can be addressed through *deep learning* [3] technologies since instead of manual feature engineering, deep learning models (e.g., *Convolutional Neural Network* (CNN)) are characterized by the automated feature learning and classifier building, which lead to its extremely efficient performance. The recent success of deep learning techniques in natural language processing, computer vision, and speech processing applications motivates us to use such techniques for universal image forensic detection as this area is still unexplored yet promising.

- With ever-increasing rate of cybercrimes and intelligent forgeries in the digital world, it would not take much to render the present digital forensic state-of-the-art, obsolete. With continuous effort by adversaries to break the existing forensic tools, an equally vigorous effort is mandatory to produce more and more efficient forensic techniques. In this respect, an emerging field is that of anti-forensics or counter-forensics [2]. Counter-forensics is nothing but countermeasures to forensic analyzes. The set of technologies adopted by adversaries to hide their identities from forensic analyzers or compromise legal evidences, hence to mislead forensic analysis, is collectively referred to as counter-forensics. Detailed investigation of counter-forensic technologies such as source counterfeiting, tamper hiding, synthesis of traces of authenticity, etc., is extremely important to cope with the ever-increasing digital forgeries and cybercrimes.
- Theoretical analysis of the limits of image forensic should be further investigated.
- Applicability of image forensics techniques for video forensics also needs to be explored.

Hence, we conclude that digital image forensics is an exciting field of research with plenty of open problems and is certain to see intensive research activities in the near future.

## References

1. Bayar, B., Stamm, M.C.: A deep learning approach to universal image manipulation detection using a new convolutional layer. In: Proceedings of the 4th ACM Workshop on Information Hiding and Multimedia Security, pp. 5–10. ACM (2016)
2. Böhme, R., Kirchner, M.: Counter-forensics: attacking image forensics. In: Digital Image Forensics, pp. 327–366. Springer (2013)

3. Bondi, L., Baroffio, L., Güera, D., Bestagini, P., Delp, E.J., Tubaro, S.: First steps toward camera model identification with convolutional neural networks. IEEE Signal Process. Lett. **24**(3), 259–263 (2017)
4. Rao, Y., Ni, J.: A deep learning approach to detection of splicing and copy-move forgeries in images. In: 2016 IEEE International Workshop on Information Forensics and Security (WIFS), pp. 1–6. IEEE (2016)

MIX
Papier aus verantwortungsvollen Quellen
Paper from responsible sources
FSC® C105338

If you have any concerns about our products,
you can contact us on
**ProductSafety@springernature.com**

In case Publisher is established outside the EU,
the EU authorized representative is:
**Springer Nature Customer Service Center GmbH**
**Europaplatz 3, 69115 Heidelberg, Germany**

Printed by Libri Plureos GmbH
in Hamburg, Germany